T0073110

Ecological Research Monographs

Series Editor

Yoh Iwasa, Department of Bioscience, Kwansei-Gakuin University, Hyogo, Japan

More information about this series at http://www.springer.com/series/8852

Hitoshi Sakio

Editor

Long-Term Ecosystem Changes in Riparian Forests

Monitoring Sites 1000
Since 2003

Springer Open

Editor
Hitoshi Sakio
Sado Island Center for Ecological
Sustainability
Niigata University
Niigata, Japan

ISSN 2191-0707   ISSN 2191-0715   (electronic)
Ecological Research Monographs
ISBN 978-981-15-3008-1   ISBN 978-981-15-3009-8   (eBook)
https://doi.org/10.1007/978-981-15-3009-8

This Springer imprint is published by the registered company Springer Nature Singapore Pte Ltd.
The registered company address is: 152 Beach Road, #21-01/04 Gateway East, Singapore 189721, Singapore

Preface

This book represents the results of more than 30 years of long-term ecological research (LTER) in riparian forest ecosystems. Considerable LTER has been conducted, and I want more people to recognize its importance and support it in the future. Hence, I will make this book available open access to the world.

LTER on forests is important, both to clarify the life history characteristics of the constituent tree species of forests and to observe forest changes directly. LTER can also capture phenomena that could not have been predicted at the beginning of the study, such as large but infrequent natural disturbances and slow phenological changes resulting from climate change, and is also important for clarifying their effects. LTER in Japan began in the late 1980s, and the Monitoring Sites 1000 Project of the Ministry of the Environment of Japan began in earnest this century.

Riparian forests not only play important roles in riverine ecosystems through their ecological functions but also provide biodiversity and ecosystem services. However, clear-cutting operations in mountain regions after World War II led to the loss of natural riparian forests and their functions. Under these circumstances, the remaining natural riparian forests are both scientifically valuable and important as a model of regeneration and restoration and a source of genetic resources.

The Ooyamazawa riparian forest is one of the remaining old-growth natural forests in Japan. Many of the trees are over 200 years old, and the forest has very high plant species diversity. A feasibility study was initiated in Ooyamazawa in 1983, and full-scale LTER began there in 1987. Initially, the research focused on the life history of the forest trees. Subsequently, it has expanded to include the coexistence of trees and forest floor herbs, and since 2008 studies of ground beetles and avifauna have been in progress.

One long-term study captured the impact of Sika deer, whose population has increased since 2000, on the riparian forest ecosystem. A nearly 30-year survey of flowering and fruiting revealed that climate change has affected the reproductive characteristics of *Fraxinus platypoda* trees. These studies have demonstrated the effectiveness and importance of long-term studies of forest dynamics, and continued monitoring can be expected to produce new research results.

These studies of the Ooyamazawa riparian forest have received support from various funding sources. The Monitoring Sites 1000 Project of the Ministry of the Environment of Japan has supported this research since 2008. This research has also been supported by JSPS KAKENHI grants JP20380091, JP25252029, and JP25450209.

The government of Saitama Prefecture has provided facilities, and Saitama Forest Science Museum has provided on-site support for the research. The incorporated nonprofit organization *"Mori to Mizu no Genryu bunkajuku"* has also continued to support our research. I thank the authors for submitting their manuscripts and all the scientists and researchers who participated in these studies. Finally, I thank Springer and its staff for their assistance and encouragement.

Niigata, Japan   Hitoshi Sakio

Contents

Part IV   Ecosystem Changes in Riparian Forests

Part V   Conclusion

Part I
Introduction

Chapter 1
The Ooyamazawa Riparian Forest: Introduction and Overview

Hitoshi Sakio

Abstract Long-term ecological research (LTER) began at the Ooyamazawa riparian forest research site in 1983. Ooyamazawa comprises a representative cool-
temperate zone old-growth riparian forest that is species-rich and contains complex
topography. The Ooyamazawa river basin comprises at least 230 species of vascular
plants, including 46 woody tree species that are found within the research site.
Researchers have used this site to study forest structure and tree life histories over
a 35-year period. In particular, research has focused on the life histories of the
dominant canopy species *Fraxinus platypoda*, *Pterocarya rhoifolia*, and
Cercidiphyllum japonicum. After the research site was registered as a Core Site of
the Monitoring Sites 1000 Project, research began on avifauna and ground beetles, in
addition to ongoing forest research.

Keywords Climate characteristics · Historical research site · Long-term ecological
research · Natural disturbance · Old-growth forest · Ooyamazawa riparian forest ·
Riparian vegetation · Study design · Topography

1.1 Introduction

The Ooyamazawa riparian forest is a representative riparian forest in a cool-
temperate zone in Japan. Many researchers have visited the forest to conduct studies,
and numerous papers (Sakio 1993, 1997; Sakio et al. 2002, 2013; Kubo et al. 2000,
2001a, b, 2005, 2008, 2010; Kawanishi et al. 2004, 2006; Sato et al. 2006) and books
(Sakio and Tamura 2008; Sakio 2017) have been published. This research site
attracts attention not only as a riparian forest in Japan but also due to its status as a
long-term ecological research (LTER) site. Ecological research at the site began in
1983 and has been ongoing ever since. During this period, substantial data on tree

H. Sakio (✉)
Sado Island Center for Ecological Sustainability, Niigata University, Niigata, Japan
e-mail: sakio@agr.niigata-u.ac.jp; sakiohit@gmail.com

© The Author(s) 2020   3
H. Sakio (ed.), *Long-Term Ecosystem Changes in Riparian Forests*, Ecological
Research Monographs, https://doi.org/10.1007/978-981-15-3009-8_1

life histories and coexistence have accumulated. Major changes have occurred in the forest; for example, forest floor vegetation has decreased dramatically as deer populations have increased. The reproductive characteristics of trees have also shown long-term changes, possibly due to global warming.

1.2   History of Long-Term Research in the Ooyamazawa Riparian Forest

The Ooyamazawa riparian forest is part of the forest land owned by Saitama Prefecture. It is located in Ooyamazawa, Nakatsugawa, Chichibu, Saitama, Japan (Fig. 1.1). The forest was donated to Saitama Prefecture in 1930 by Seiroku Honda, PhD. The Saitama Prefecture forest is made up of 12 forest stands, including old-growth forest, coppice forest, and plantations of *Cryptomeria japonica* and *Chamaecyparis obtusa*. Ooyamazawa comprises the 106th and 107th forest stands.

Wood production by selective cutting of old-growth forest was conducted in the downstream part of Ooyamazawa from 1936 to 1941, and there are no logging records prior to 1936. Timbers were transported downstream by water discharged from a log dam known as a "*Teppou zeki*" (Fig. 1.2). At that time, the upper basin of Ooyamazawa had not been cut down and the natural forest was conserved.

Construction of a forest road along the Ooyamazawa stream began in 1964 through this forest stand (Fig. 1.3). However, public opposition to the felling of the national forest of the Chichibu Mountains increased, such that construction of the forest road was stopped mid-way through 1969. Most of the natural riparian forests in mountain regions in Japan have been lost to clear cutting, and were replaced by conifer plantations after World War II; thus, there are few natural riparian forests left. The upper part of the Ooyamazawa riparian forest is valuable because it has not been affected by human activities such as logging or erosion control works.

In 1950, this area was designated as a Class II special zone (i.e., an area in which agriculture, forestry, and fishery activities must be coordinated to suit the environment) of Chichibu-Tama National Park. In 2000, the area was renamed Chichibu-Tama-Kai National Park.

In 2013, a portion of these stands (109.12 ha), including the riparian zone (5 ha), was designated as a Natural Monument of Saitama Prefecture. This area was designated as the Kobushi Biosphere Reserve in 2019.

In October 1983, a riparian area research plot (core plot: 60 m × 90 m) was demarcated, and was subsequently extended to 4.71 ha along the mountain stream from 1991 to 1998. In December 2006, the research site was registered as an associate site of the Japan LTER Network (JaLTER; http://www.jalter.org/en/researchsites/). In 2008, a 1-ha plot, including a 0.54-ha core plot in the research site (Fig. 1.4), was also registered as a Core Site of the Monitoring Sites 1000 Project by the Ministry of the Environment, Japan (http://www.biodic.go.jp/moni1000/index.html).

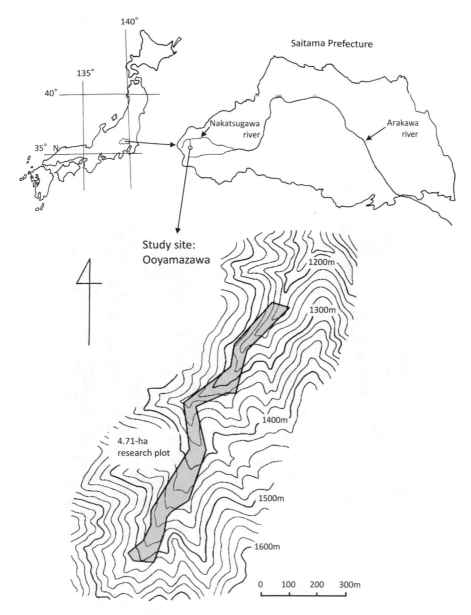

Fig. 1.1 Map of the experimental area within the research site

1.3   Site Description

The Ooyamazawa riparian forest research site (35°57′48″ N, 138°45′22″ E) is located in the Chichibu Mountains of the Kanto region of central Japan (Fig. 1.1). This site is located in a riparian zone along a small stream (Ooyamazawa) of the

Fig. 1.2 A log dam, known as a *Teppou zeki*, in the Ooyamazawa stream. Photograph courtesy of Saitama Prefecture

Nakatsugawa River branch of the Arakawa River in Saitama Prefecture, central Japan. The site is situated within the Chichibu-Tama-Kai National Park and ranges from 1210 to 1530 m above sea level (a.s.l.). The 106th and 107th forest stands in Ooyamazawa cover approximately 512 ha.

Fig. 1.3   The entrance of a forest road along the Ooyamazawa stream

1.4   Climate Characteristics in the Ooyamazawa Riparian Forest

There is a large difference in climate between the Pacific Ocean and Japan Sea sides. On the Japan Sea side, snow and rain are abundant in the winter due to seasonal northwest winds. On the Pacific side, a significant amount of rain falls during the summer due to southeast seasonal winds blowing from the Pacific Ocean. The Ooyamazawa research site is on the Pacific Ocean side. Japanese forest zones are divided into four categories: subarctic forest, cool-temperate forest, warm temperate forest, and subtropical forest; Ooyamazawa is a cool-temperate forest.

1.4.1   Air Temperature

The mean annual temperature at the study site (1450 m a.s.l.) is 7.1 °C. The mean air temperature is 18.3 °C in the warmest month (August) and −5.2 °C in the coldest month (January) (Fig. 1.5). The monthly average maximum temperature is 20.3 °C in August and the monthly average minimum temperature is −8.6 °C in January

Fig. 1.4 Core Site of the Monitoring Sites 1000 Project, as designated by the Ministry of the Environment, in the Ooyamazawa riparian forest

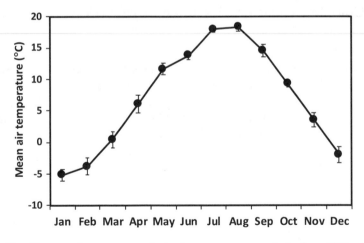

Fig. 1.5 Monthly mean air temperatures in the Ooyamazawa research site from 2008 to 2018

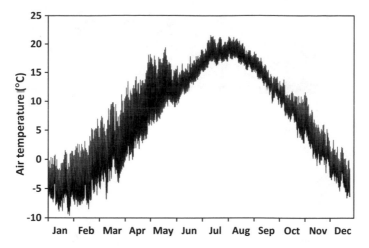

Fig. 1.6 Diurnal ranges of air temperature in the Ooyamazawa research site from 2008 to 2018

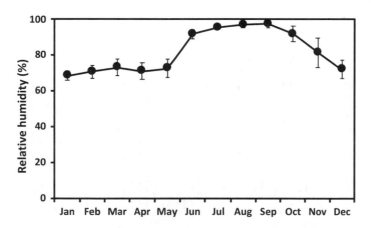

Fig. 1.7 Monthly mean relative air humidity in the Ooyamazawa research site from 2014 to 2018

(2008–2018). The diurnal ranges of air temperature from January to May are larger than in other months due to the development of canopy leaves (Fig. 1.6).

1.4.2 Relative Air Humidity

The mean annual air humidity was 81.9% from 2014 to 2018. The maximum mean air humidity was 97.4% in September and the minimum was 68.3% in January (Fig. 1.7). The air humidity is higher in the summer season than in the winter season. The diurnal ranges of air humidity were larger from January to May than in other months (Fig. 1.8).

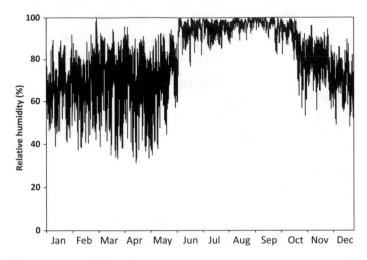

Fig. 1.8 Diurnal ranges of relative air humidity in the Ooyamazawa research site from 2014 to 2018

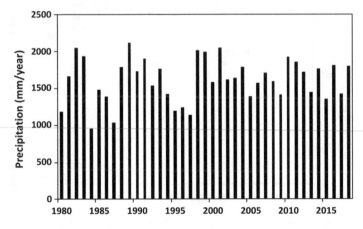

Fig. 1.9 Changes in precipitation measured at the Mitsumine automated weather station in Saitama Prefecture from 1980 to 2018. This station is 975 m above sea level (a.s.l.) and 15 km from the Ooyamazawa research site

1.4.3 Precipitation

The mean annual precipitation at the Mitsumine automated weather station in Saitama Prefecture, located 15 km from the research site at 975 m a.s.l., was 1611.7 mm from 1980 to 2018 (Fig. 1.9). Precipitation is higher in summer than in winter due to the rainy season and the occurrence of typhoons (Fig. 1.10). In particular, due to a large typhoon, 717.5 mm of precipitation was recorded in August 2016. On 14 August 1999, we recorded 440 mm of daily precipitation. The mean

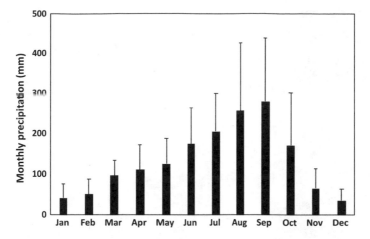

Fig. 1.10 Monthly mean precipitation at the Mitsumine automated weather station in Saitama Prefecture from 1980 to 2018

maximum snow depth at the research site was approximately 30 cm between January and March. However, in 2014, there were heavy snowfalls and the snow depth was estimated to exceed 3 m in the valley bottom due to avalanches.

1.5   Geology, Topography, Natural Disturbances, and Soils

The study area is covered by graywacke and sandstone of the Paleozoic era. The Chichibu Mountains have a complex topography with steep slopes (greater than 30°) and a network of mountain streams. This area is registered as a Japanese Geopark.

The topography of the Ooyamazawa riparian research plot along the stream (1.2 km) can be divided into two parts: the downstream area is a V-shaped valley with a steep slope of 30° (Fig. 1.11), and the upstream area is a wide floodplain characterized by debris flows and landslides with a slope of 12° (Fig. 1.12).

In the Chichibu Mountains, large typhoons accompanied by >300 mm of diurnal precipitation have occurred once every decade throughout the twentieth century (Saitama Prefecture & Kumagaya Local Meteorological Observatory 1970). These heavy rains result in debris flows (Fig. 1.13), surface landslides, and channel movements that do not improve light conditions on the mountains, whereas another type of disturbance, involving the destruction of large areas of canopy trees, improves light conditions. For example, large earthquakes and typhoons can cause large mass movements through landslides (Fig. 1.14). Sedimentation and erosion of

Fig. 1.11 The V-shaped valley in the downstream area. The hillside is very steep with a slope of >30°

sand and gravel occur on a small-scale every year in the active channels during the rainy and typhoon seasons. Flooding during these times has resulted in the emergence of abandoned channels and large deposits (Sakio 1997).

A typical soil for this area is a moderately moist brown forest soil (BD). However, riparian zones have complex microtopography with many soil types. The substratum of the active channel is sand and/or gravel, while that of the hillslope is mature soil. In the active channel, the ground surface is covered with large rocks, gravel, and sand. In the winter season, only groundwater flow is present. In the abandoned channel, no movement of sand or gravel occurs due to stream flow and there is a dense *Fraxinus platypoda* and *Pterocarya rhoifolia* sapling bank. In the floodplain, there are two A horizon layers containing plant roots due to repeated sedimentation. Meanwhile, on the hillslope, a thick litter layer and a humus layer can be observed in the soil profile (Sakio 1997; Fig. 1.15).

Fig. 1.12 The wide floodplain characterized by debris flows and landslides in the upstream area

1.6 Vegetation

The Ooyamazawa riparian forest (4.71 ha) is a species-rich natural forest. This forest is situated in the upper areas of a cool-temperate, deciduous broad-leaved forest zone that ranges from 700 to 1600 m a.s.l; it is a typical mountain region riparian forest classified as a *Dryopterido—Fraxinetum commemoralis* type (Maeda and Yoshioka 1952) originally reported by Suzuki (1949). However, Ohno (2008) suggested that it should be classified as *Cacalio yatabei—Pterocaryetum rhoifoliae*, as per the riparian forest communities in the mountain belts of the inland Chubu and Kanto regions.

The Ooyamazawa river basin contains at least 230 species of vascular plants (Kawanishi et al. 2006). Forty-six woody tree species were reported in a 4.71-ha research plot (Sakio 2008); of these, nineteen were canopy tree species. Canopy layer species are over 30 m in height and include *F. platypoda*, *P. rhoifolia*, and *Cercidiphyllum japonicum*. The dominant plants in the subcanopy layer are *Acer shirasawanum* and *Acer pictum*, while the lower layer is dominated by *Acer carpinifolium* and *Acer argutum*. The main herbaceous species in the understory are *Mitella pauciflora*, *Asarum caulescens*, *Meehania urticifolia*, and *Dryopteris polylepis*.

Fig. 1.13 Debris flow after heavy rain (Sakio 2008)

1.7 History of the Ground Design of the Research Plot

We demarcated a permanent plot of 60 × 90 m (0.54 ha), in the Ooyamazawa riparian forest in October 1983 (Fig. 1.16), and researched the forest structure and soil profile (Sakio 1997). This plot was extended to 4.71 ha along the mountain stream from 1991 to 1998. The plot was long (1170 m) and narrow (30–60 m), and comprised 30 × 30-m subplots that covered the lower part of a hillside adjacent to the riparian area (Sakio et al. 2002).

Fig. 1.14 A deep-seated landslide after a typhoon, which caused large mass movements and changes in light conditions

All living trees ≥4 cm diameter at breast height (DBH) were numbered and identified to the species level, and divided into canopy trees, subcanopy trees (DBH ≥ 10 cm), and small young trees (DBH < 10 cm). Canopy trees reached the top stratum of the forest. All trees were mapped using a compass survey.

Seed and leaf production have been measured from 1987 to the present. We set 20 conical litter traps approximately 1 m above the ground in a regular pattern within the core plot (0.54 ha) in May 1987 (Fig. 1.17). Litter traps were made of nylon netting (~1.0-mm mesh) with 0.5 m radius openings. When the plot became a Monitoring Sites 1000 Project Core Site (1 ha, including the 0.54 ha core plot) in April 2008, we added five new litter traps, giving 25 traps in total (Fig. 1.17). The litter trap size was changed to a 0.4-m radius in 2010. We collected the contents of the litter traps and brought them back to the laboratory every month between May and November. In the winter, the traps were left in the plot. In the laboratory, we divided the litter into seeds, flowers, fruits, fallen leaves, and branches/bark, measured the weight of each litter type and counted the seeds. From 1987 to 1994,

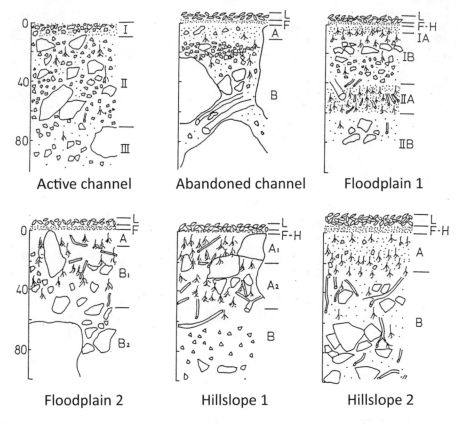

Fig. 1.15 Soil profiles at the core plot in the Ooyamazawa research site (Sakio 1997). The soil of the active channel was immature and divided into three layers. Floodplain 1 showed two A horizons due to repeated sedimentation

only *F. platypoda* was measured; after 1995, *F. platypoda*, *P. rhoifolia*, *C. japonicum*, *Tilia japonica*, *Ulmus laciniata*, *Carpinus cordata*, *Abies homolepis*, and some *Acer* species were also measured.

Studies of ground beetles have been in progress at the Monitoring Sites 1000 Project Core Site since 2008. Five subplots (5 m × 5 m) have been established within the Core Site and four pitfall traps were installed in each subplot. In addition, a bird survey project conducted at five fixed points in Ooyamazawa was established in 2010.

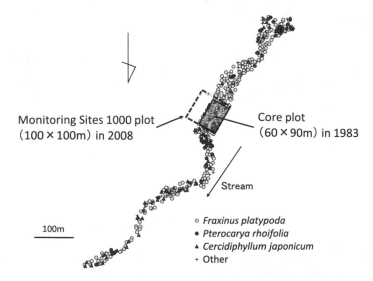

Fig. 1.16 History of the study design of the Ooyamazawa research site. The core plot (60 m × 90 m) was demarcated in 1983. The research site was extended to 4.71 ha along the mountain stream from 1991 to 1998. The core plot was extended to a 1-ha plot and registered as a Core Site of the Monitoring Sites 1000 Project in 2008

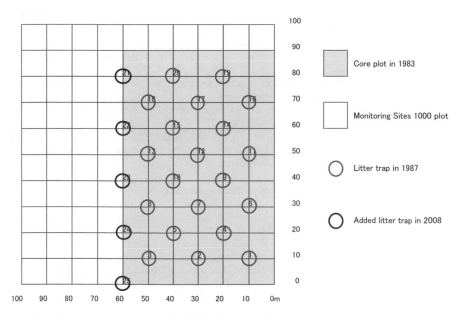

Fig. 1.17 Layout of litter traps in the Monitoring Sites 1000 Project Core Site

References

Kawanishi M, Sakio H, Ohno K (2004) Forest floor vegetation of *Fraxinus platypoda - Pterocarya rhoifolia* forest along Ooyamazawa valley in Chichibu, Kanto District, Japan, with a special reference to ground disturbance. J Veg Sci 21(1):15–26 (in Japanese with English summary)

Kawanishi M, Sakio H, Kubo M, Shimano K, Ohno K (2006) Effect of micro-landforms on forest vegetation differentiation and life-form diversity in the Chichibu Mountains, Kanto District, Japan. J Veg Sci 23:13–24

Kubo M, Shimano K, Sakio H, Ohno K (2000) Germination sites and establishment conditions of *Cercidiphyllum japonicum* seedlings in the riparian forest. J Jpn For Soc 82:349–354 (in Japanese with English summary)

Kubo M, Shimano K, Sakio H, Ohno K (2001a) Sprout trait of *Cercidiphyllum japonicum* based on the relationship between topographies and sprout structure. J Jpn For Soc 83(4):271–278 (in Japanese with English summary)

Kubo M, Shimano K, Ohno K, Sakio H (2001b) Relationship between habitat of dominant trees and vegetation units in Chichibu Ohyamasawa riparian forest. Veg Sci 18(2):75–85 (in Japanese with English summary)

Kubo M, Sakio H, Shimano K, Ohno K (2005) Age structure and dynamics of *Cercidiphyllum japonicum* sprouts based on growth ring analysis. Forest Ecol Manag 213:253–260

Kubo M, Kawanishi M, Shimano K, Sakio H, Ohno K (2008) The species composition of soil seed banks in the Ooyamazawa riparian forest, in the Chichibu Mountains, central Japan. J Jpn For Soc 90(2):121–124 (in Japanese with English summary)

Kubo M, Shimano K, Sakio H, Isagi Y, Ohno K (2010) Difference between sprouting traits of *Cercidiphyllum japonicum* and *C. magnificum*. J Forest Res 15(5):337–340

Maeda T, Yoshioka J (1952) Studies on the vegetation of Chichibu Mountain forest. (2) The plant communities of the temperate mountain zone. Bull Tokyo Univ For 42:129–150+3pls (in Japanese with English summary)

Ohno K (2008) Vegetation-geographic evaluation of the syntaxonomic system of valley-bottom forests occurring in the cool-temperate zone of the Japanese Archipelago. In: Sakio H, Tamura T (eds) Ecology of Riparian forests in Japan: disturbance, life history, and regeneration. Springer-Verlag, pp 49–72

Saitama Prefecture & Kumagaya Local Meteorological Observatory (1970) Weather disaster of Saitama Prefecture. Saitama Prefecture & Kumagaya Local Meteorological Observatory. Saitama (in Japanese)

Sakio H (1993) Sapling growth patterns in *Fraxinus platypoda* and *Pterocarya rhoifolia*. Jpn J Ecol 43(3):163–167 (in Japanese with English summary)

Sakio H (1997) Effects of natural disturbance on the regeneration of riparian forests in a Chichibu Mountains, central Japan. Plant Ecol 132:181–195

Sakio H (2008) Coexistence mechanisms of three riparian species in the upper basin with respect to their life histories, ecophysiology, and disturbance regimes. In: Sakio H, Tamura T (eds) Ecology of riparian forests in Japan: disturbance, life history and regeneration. Springer-Verlag, pp 75–90

Sakio H (2017) A natural history of trees in riparian forests. University of Tokyo Press, 260 pp (in Japanese)

Sakio H, Tamura T (eds) (2008) Ecology of riparian forests in Japan: disturbance, life history and regeneration. Springer-Verlag, 339 pp

Sakio H, Kubo M, Shimano K, Ohno K (2002) Coexistence of three canopy tree species in a riparian forest in the Chichibu Mountains, central Japan. Folia Geobot 37:45–61

Sakio H, Kubo M, Kawanishi M, Higa M (2013) Effects of deer feeding on forest floor vegetation in the Chichibu Mountains, Japan. J Jpn Soc Reveget Tech 39(2):226–231 (in Japanese with English summary)

Sato T, Isagi Y, Sakio H, Osumi K, Goto S (2006) Effect of gene flow on spatial genetic structure in the riparian canopy tree *Cercidiphyllum japonicum* revealed by microsatellite analysis. Heredity 96:79–84

Suzuki T (1949) Temperate forest vegetation in the upper stream area of the River Tenryu. Sylvic Techn Gijitsu-Kenkyu 1:77–91 (in Japanese with English summary)

Part II
Life History and Regeneration Processes of Riparian Woody Species

Chapter 2
Fraxinus platypoda

Hitoshi Sakio

Abstract The ash species *Fraxinus platypoda* is the dominant canopy tree species at Ooyamazawa riparian forest. I investigated flowering, seed production, germination, seedling survival and growth, and structural measures in *F. platypoda*. Flowering and seed production demonstrated a clear masting pattern over a 28-year period. The seeds of *F. platypoda* germinated in most environments, but seedling survival was regulated by microtopographic factors. I suggest that gap formation processes may be vital to the establishment of canopy-height individuals. The overall forest structure and spatial distribution of *F. platypoda* suggested that all individuals of this species occurring at Ooyamazawa regenerated simultaneously following a large-scale disturbance event 200 years ago. Since this event occurred, saplings of *F. platypoda* have regenerated in canopy gaps. It is probable that *F. platypoda* succeeds as a dominant species in riparian forests by regenerating in response to disturbance at multiple scales.

Keywords Advanced sapling · Dioecy · Disturbance regime · Flowering · Germination · Life history · Microtopography · Reproductive strategy · Seed production · Seedling

2.1 Introduction

Native riparian forests are distributed along the Ooyamazawa stream in the Chichibu Mountains of the Kanto region, central Japan. In these forests, *Fraxinus platypoda* Oliv. (*Oleaceae*) is one of the dominant canopy tree species, coexisting with *Pterocarya rhoifolia* and *Cercidiphyllum japonicum*. The Chichibu Mountains harbor a very complex topography characterized by steep slopes, with tree diversity distributed among the resulting microhabitats. For example, *Tsuga sieboldii* and

H. Sakio (✉)
Sado Island Center for Ecological Sustainability, Niigata University, Niigata, Japan
e-mail: sakio@agr.niigata-u.ac.jp; sakiohit@gmail.com

© The Author(s) 2020   23
H. Sakio (ed.), *Long-Term Ecosystem Changes in Riparian Forests*, Ecological
Research Monographs, https://doi.org/10.1007/978-981-15-3009-8_2

Chamaecyparis obtusa forests are distributed along ridges, *Fagus crenata* and *Fagus japonica* on mountain slopes, and *F. platypoda, P. rhoifolia*, and *C. japonicum* in valleys (Maeda and Yoshioka 1952; Tanaka 1985). The regeneration mechanisms of *T. sieboldii, F. crenata*, and *F. japonica* have been extensively studied by Suzuki (1979, 1980, 1981a, b), Nakashizuka and Numata (1982a, b) and Nakashizuka (1983, 1984a, b), and Ohkubo et al. (1988, 1996), respectively. These researchers examined forest regeneration within the context of gap dynamics theory. Gaps are open spaces within the forest canopy layer formed by die back, trunk breakage, and uprooted trees. In general, gaps occur on a small scale and are not often accompanied by soil disturbance, with the exception of uprooted trees.

On the other hand, disturbances within riparian zones vary in type, frequency, magnitude, and size compared to canopy gap formation on hillslopes. In steeper mountain regions, valley floor landforms are sculpted by fluvial processes and a variety of mass soil movement processes from tributaries and adjacent hillslopes (Gregory et al. 1991).

The ash species, *Fraxinus platypoda*, is a late successional species in riparian habitats of cool temperate forests distributed along the Pacific coast of Japan. This species is distributed from Tochigi Prefecture in the north to Miyazaki Prefecture in the south (Fig. 2.1). Kisanuki et al. (1992) and Ann and Oshima (1996) examined regeneration mechanisms within a gap dynamics framework in mixed forests of *F. Platypoda* and *P. rhoifolia*. However, the regeneration of *F. platypoda* may also be related to large-scale natural disturbances such as debris flows and landslides in the riparian zone. In this chapter, I present long-term research related to the life history and regeneration process of *F. platypoda* with respect to natural disturbances within the riparian zone.

2.2   Study Species

Fraxinus platypoda is a deciduous canopy species that can reach up to 40 m in height and 150 cm in diameter at breast height (DBH, 130 cm) (Fig. 2.2). This species is well adapted to stream disturbances that vary in frequency and size, and it dominates forests in riparian zones (Sakio 1997).

Branching occurs above the trunk of the tree. Most individuals have a single trunk and rarely exhibit sprouts, similar to species such as *P. rhoifolia* and *C. japonicum*. Leaves are impari-pinnate compound, consisting of 7–9 leaflets, and leaf length is about 25–35 cm and decussate-opposite (Fig. 2.3). The base of the petiole markedly bulges and holds the stem. Open hairs occur along the middle vein on the back of the leaf, but others are hairless. The apical leaflet is oblong-ellipsoid oblanceolate in shape, and is 8–20 cm in length and 3–7 cm in width, with a small petiole of 1–2 cm in length. The side leaflets lack a petiole, and the base is wedged with fine serrations. Twigs are thick, gray-brown, and hairless, and many are oval lenticel. The pith is thick. Branches grow rapidly in early spring and stop growing in June (Sakio 1993). The root system is concentrated in the shallow part of the ground surface. *Fraxinus*

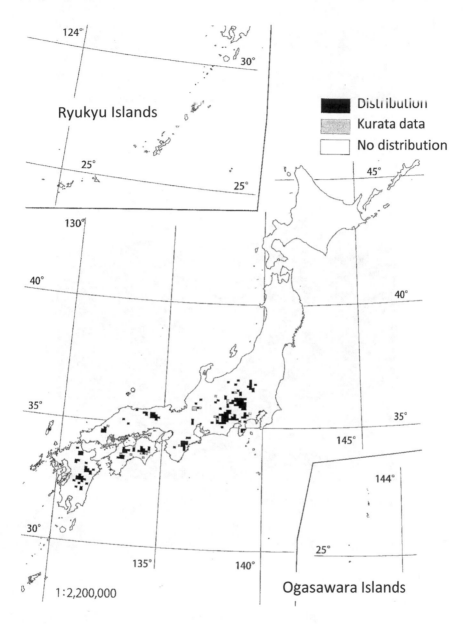

Fig. 2.1   Distribution of *F. platypoda*. Modified after Kawahara et al. (2009)

platypoda has medium to large-diameter straight roots and horizontal roots and is classified as a deep root type. Fine roots are dense, but root hairs are rare (Karizumi 1979). The species is easily uprooted by scouring running water. In saplings, sedimentation by soil and sand easily produces adventitious roots (Sakio 2002).

Fig. 2.2 *F. platypoda* forest in the Ooyamazawa riparian forest

2.3 Reproductive Traits

2.3.1 Flower

Sexual expression of *F. platypoda* is not obvious. The species exhibits two morphological types: one with male flowers and the other with hermaphrodite flowers (Fig. 2.4). Both flowers lack a perianth. Male flowers have one stamen with a pair of anthers, while hermaphrodite flowers have one pistil with a pair of anthers. Whether the male portion of hermaphrodite flowers of *F. platypoda* is functional is unknown. However, sexual expression of *F. platypoda* may be functionally considered to be androdioecy, as the pollen of hermaphrodite flowers exhibits germination ability. Because the breeding characteristics of *F. platypoda* are not clearly understood, I

Fig. 2.3   Leaves of *F. platypoda*

Hermaphrodite flower   Male flower

Fig. 2.4   Two types of flowers of *F. platypoda*

treat the sexual expression of *F. platypoda* as dioecy, i.e., female trees and male trees, in this chapter.

The timing of flowering in *F. platypoda* differs depending on altitude but, typically, occurs in mid-April (700 m a.s.l) to mid-May (1500 m a.s.l.). In the Ooyamazawa riparian forest (1500 m a.s.l.), flowering occurs from the beginning of May to mid-May. However, the exact timing fluctuates annually.

Flowering data collected over 28 years have demonstrated clear fluctuations of flower values for *F. platypoda* in the core research plot (Fig. 2.5). The numbers of female and male trees in the core plot (0.54 ha) were 26 and 20, respectively. The rank of flowering for female and male trees was scored from 1 to 5 by observation using binoculars. The average flowering rank of all individuals was 3.22 for females

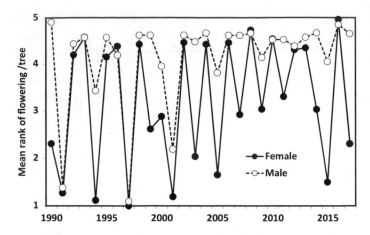

Fig. 2.5 Mean rank of flowering of female and male trees of *F. platypoda* from 1990 to 2017. The numbers of female and male trees were 26 and 20, respectively

and 4.11 for males; values were significantly higher in males. The coefficient of variation (CV) was 0.48 for females and 0.28 for males; values were significantly higher in females. In this forest, although the flowering interval was 2–3 years, the interval changed over the 28 years. The flowering of the two sexes exhibited clear synchronization until 2002, but synchronization ceased thereafter. After 2002, more males have flowered every year, while females have retained a distinct interval. The change in flowering fluctuation after 2002 may have been driven by external factors, such as climate warming.

2.3.2 Seed Production

After flowering, fruit (samara) of *F. platypoda* continues to grow, causing abortion. The pericarp finishes growing at the end of August, at which point seeds begin to grow. Oven-dried matured fruit of *F. platypoda* weighed 144 ± 24 mg, and the dry weight of seeds was 80 ± 17 mg (Fig. 2.6; Sakio et al. 2002). Seeds mature in mid-October and are dispersed by wind and stream water in November (Fig. 2.7). Many mature seeds occur in mast years, but a large number of seeds are empty or insect-damaged during non-mast years.

Clear fluctuations in seed production and flowering of *F. platypoda* occurred over the 28 years (Fig. 2.8). Because seed production was strongly positively correlated with the extent of flowering, the former is presumed to be regulated by the latter (Fig. 2.9).

Fig. 2.6 Fruits and seeds of *F. platypoda* in autumn

2.4   Germination

The seeds of *F. platypoda* germinate from the end of June to mid-July. Current-year seedlings are found on litter, gravel, mineral soil, and fallen logs, except after non-mast years. As long as the light environment is not very strong, current-year seedlings of *F. platypoda* only have cotyledons (Fig. 2.10), and the true leaf does not expand during the germination year. The germination site of *F. platypoda* is not strongly restricted by the soil and light environment, unlike *C. japonicum*, whose germination sites are limited. *F. platypoda* does not exhibit seed dormancy. The seeds of *F. platypoda* produced in autumn germinate in the early summer of the following year and do not germinate thereafter.

Fig. 2.7 Fruits of *F. platypoda* in the stream in autumn (Sakio 2008)

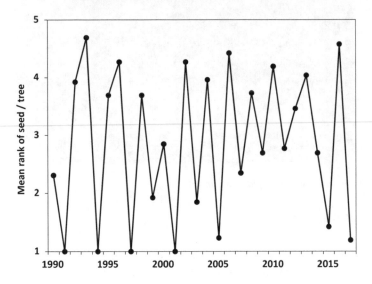

Fig. 2.8 Mean rank of seed production of *F. platypoda* from 1990 to 2017

2.5 Seedling Survival

The seeds of *F. platypoda* are randomly dispersed. Therefore, the distribution of current-year seedlings of *F. platypoda* is not affected by microenvironments such as soil and light conditions. However, the distribution pattern of seedlings of *F. platypoda* changes with the growth of seedlings, shifting from random or uniform

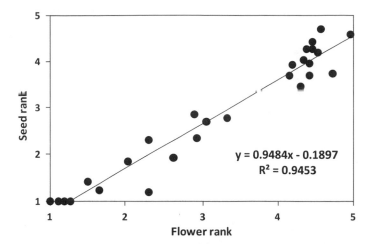

Fig. 2.9 Relationship between flower and seed rank

Fig. 2.10 Current seedling of *F. platypoda*

distribution to aggregated throughout growth (Fig. 2.11). Small-sized seedlings (height < 20 cm) tend to be distributed around the active channel (Figs. 2.11 and 2.12), while the distribution of larger seedlings (20 cm ≤ height < 1 m) is more closely related to the microtopography than to canopy gaps. These larger seedlings tend to be aggregated in abandoned channels. On hillslopes where the forest floor vegetation is dense, *F. platypoda* seedlings disappear after several years due to the effects of shade. The mean longevity of *F. platypoda* seedlings in various environments is 1.19 ± 0.58 years (Sakio et al. 2002). On the other hand, seedlings near

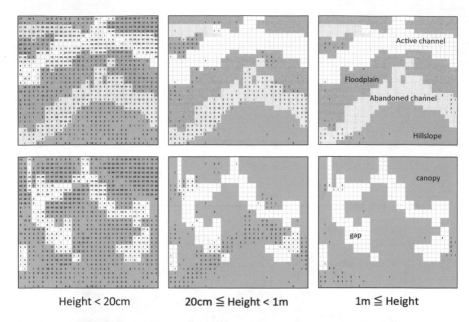

Height < 20cm   20cm ≦ Height < 1m   1m ≦ Height

Fig. 2.11 Distribution of *F. platypoda* seedlings. Upper and lower graphs show the relationship with microtopography and canopy gaps, respectively. The six plots in the figure are each 30 × 32 m, and numbers indicate the number of individuals in 1 m²

Fig. 2.12 Advanced sapling communities of *F. platypoda* along the stream

active channels exhibit long life spans and high density due to the absence of forest floor vegetation (Fig. 2.11). However, seedling communities can be destroyed by flooding, often caused by large typhoons. Gravel deposits formed during flooding events can serve as new seedling establishment sites. When the seedling community stabilizes due to channel fluctuations, it continues to grow into a large-sized seedling community (20 cm ≤ height).

In other words, the distribution of *F. platypoda* seedlings is regulated by the microtopographic variation in forest floor vegetation. In riparian forests, the light environment depends not only on the presence of canopy gaps but also on gaps in the forest floor vegetation due to stream disturbance. The dynamics of *F. platypoda* seedlings are thought to be strongly influenced by the latter.

2.6   Seedling Growth

Large-sized seedlings (1 m ≤ height) that have established on stable sites exhibit variation in growth rates depending on the light environment. New shoots of *F. platypoda* begin to elongate rapidly in early spring and stop growing in June (Sakio 1993). Seedlings under canopy gaps grow faster than seedlings under the canopy (Fig. 2.13) and have more leaves. In many forests, canopy gaps appear to be necessary for the growth of canopy trees (Suzuki 1980, 1981a; Nakashizuka and Numata 1982a, b; Nakashizuka 1983, 1984a). Therefore, even in riparian forests, gap formation via the death of canopy trees may be necessary for seedlings of *F. platypoda* to grow into the canopy.

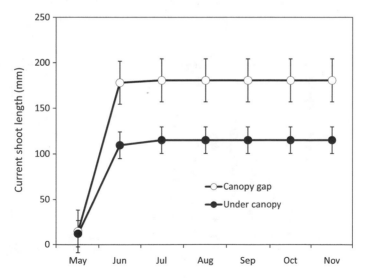

Fig. 2.13 Seasonal changes of current shoot lengths of *F. platypoda* saplings in canopy gaps and under the canopy. Modified after Sakio (1993)

2.7 Forest Structure and Spatial Distribution

In the 4.71-ha study plot within the Ooyamazawa riparian forest, we observed
418 *F. platypoda* individuals out of a total of 2214 trees (4 cm ≤ DBH) (Sakio
et al. 2002). Canopy trees, subcanopy trees (10 cm ≤ DBH), and the shrub layer
(DBH < 10 cm) accounted for 304, 59, and 55 individuals, respectively. The mean
DBH of *F. platypoda* canopy trees was 56.9 ± 19.0 cm, with a maximum of
140.5 cm (Sakio et al. 2002). The DBH distribution of *F. platypoda* was continuous
from saplings to large canopy trees (Fig. 2.14). *F. platypoda* exhibited two peaks in
DBH distribution: one formed by small trees (DBH < 10 cm) and the other formed
by the 40-cm DBH class. These data suggest that *F. platypoda* maintains sapling
banks. The peak of the 40-cm class suggests synchronous regeneration caused by a
large-scale disturbance. The relative density of dominant canopy trees of
F. platypoda is high and does not exhibit distinct fluctuations along the stream.

In the core plot (60 × 90 m), one peak of *F. platypoda* individuals occurred
within the 40–60-cm DBH class, similar to the pattern observed in the 4.71-ha plot
(Sakio 1997). In addition, increment cores of all *F. platypoda* individuals larger than
4-cm DBH were obtained using an increment borer in November 1998, in the core
plot. The age distribution of *F. platypoda* individuals was continuous from saplings
to older canopy-aged trees (Sakio 1997). Figure 2.15 presents the spatial distribution
of tree age for *F. platypoda* in the core plot. The age of most individuals was
aggregated around 200 years, but several young aggregated groups also occurred
(e.g., small patches A, C, and F). These results suggest that a large-scale disturbance
occurred around 200 years ago in the Ooyamazawa riparian forest and that
F. platypoda regenerated simultaneously. Since that time, *F. platypoda* advanced
saplings have regenerated under canopy gaps. Thus, *F. platypoda* is likely to become

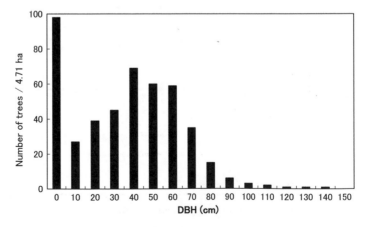

Fig. 2.14 DBH distribution of *F. platypoda* individuals in the Ooyamazawa riparian plot (4.71 ha).
Only individuals over 4 cm in DBH were measured. Modified after Sakio (2008)

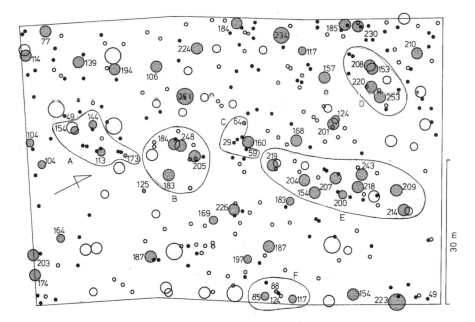

Fig. 2.15 Spatial distribution of *F. platypoda* individuals in the core plot (60 × 90 m). Numbers show the age of *F. platypoda* individuals. The size of circles reflects the DBH. Gray and black circles show *F. platypoda* individuals, and open circles show other species

an overwhelmingly dominant species within the basin because it can regenerate within sites of large-scale disturbance as well as within small gaps.

2.8   Conclusion

Fraxinus platypoda produces a large number of seeds once every few years, thus forming young advanced sapling communities within various microtopographic habitats. In particular, advanced sapling communities are formed in gravel deposits along mountain streams. These sapling communities continue to be regenerated after repeated destruction by mountain stream disturbances and subsequent regeneration on new gravel deposits. After large disturbances, *F. platypoda* regenerates within all river basins and also fills canopy gaps by advanced saplings when small gaps are formed. Thus, *F. platypoda* succeeds as a dominant species in riparian forests by regenerating in response to various scales of disturbances throughout its life history.

References

Ann SW, Oshima Y (1996) Structure and regeneration of *Fraxinus spaethiana* - *Pterocarya rhoifolia* forests in unstable valleys in the Chichibu Mountains, central Japan. Ecol Res 11 (3):363–370

Gregory SV, Swanson FJ, Mckee WA, Cummins KW (1991) An ecosystem perspective of riparian zones: focus on links between land and water. BioScience 41:540–551

Karizumi N (1979) Illustration of tree roots. Seibundo Shinkousha (in Japanese)

Kawahara T, Watanabe S, Matsui T, Takahashi M (2009) Silvics of Japan Bunpuzu. In: The publishing association of Silvics of Japan (ed) Silvics of Japan I. Japan Forestry Investigation Committee, Tokyo, pp 725–760

Kisanuki H, Kaji M, Suzuki K (1992) Structure and regeneration process of ash (*Fraxinus spaethiana* Ling.) stands in Chichibu Mountains. Bull Tokyo Univ For 88:15–32 (in Japanese with English summary)

Maeda T, Yoshioka J (1952) Studies on the vegetation of Chichibu Mountain forest. (2) The plant communities of the temperate mountain zone. Bull Tokyo Univ For 42:129–150+3pls (in Japanese with English summary)

Nakashizuka T (1983) Regeneration process of climax beech (*Fagus crenata* Blume) forests. III. Structure and development processes of sapling populations in different aged gaps. Jpn J Ecol 33:409–418

Nakashizuka T (1984a) Regeneration process of climax beech (*Fagus crenata* Blume) forests IV. Gap formation. Jpn J Ecol 34:75–85

Nakashizuka T (1984b) Regeneration process of climax beech (*Fagus crenata* Blume) forests. V. Population dynamics of beech in a regeneration process. Jpn J Ecol 34:411–419

Nakashizuka T, Numata M (1982a) Regeneration process of climax beech forests I. Structure of a beech forest with the undergrowth of *Sasa*. Jpn J Ecol 32:57–67

Nakashizuka T, Numata M (1982b) Regeneration process of climax beech forests II. Structure of a forest under the influences of grazing. Jpn J Ecol 32:473–482

Ohkubo T, Kaji M, Hamaya T (1988) Structure of primary Japanese beech (*Fagus japonica* Maxim.) forest in the Chichibu Mountains, central Japan, with special reference to regeneration processes. Ecol Res 3:101–116

Ohkubo T, Tanimoto T, Peters R (1996) Response of Japanese beech (*Fagus japonica* Maxim.) sprouts to canopy gaps. Vegetatio 124:1–8

Sakio H (1993) Sapling growth patterns in *Fraxinus platypoda* and *Pterocarya rhoifolia*. Jpn J Ecol 43(3):163–167 (in Japanese with English summary)

Sakio H (1997) Effects of natural disturbance on the regeneration of riparian forests in a Chichibu Mountains, central Japan. Plant Ecol 132:181–195

Sakio H (2002) Survival and growth of planted trees in relation to debris movement on gravel deposit of a check dam. J Jpn For Soc 84(1):26–32 (in Japanese with English summary)

Sakio H (2008) Coexistence mechanisms of three riparian species in the upper basin with respect to their life histories, ecophysiology, and disturbance regimes. In: Sakio H, Tamura T (eds) Ecology of riparian forests in Japan: disturbance, life history and regeneration. Springer, pp 75–90

Sakio H, Kubo M, Shimano K, Ohno K (2002) Coexistence of three canopy tree species in a riparian forest in the Chichibu Mountains, central Japan. Folia Geobot 37:45–61

Suzuki E (1979) Regeneration of *Tsuga sieboldii* forest. I. Dynamics of development of a mature stand revealed by stem analysis data. Jpn J Ecol 29:375–386 (in Japanese with English Synopsis)

Suzuki E (1980) Regeneration of *Tsuga sieboldii* forest. II. Two cases of regenerations occurred about 260 and 50 years ago. Jpn J Ecol 30:333–346 (in Japanese with English Synopsis)

Suzuki E (1981a) Regeneration of *Tsuga sieboldii* forest. III. Regeneration under a canopy gap with low density of conifer seedlings and a method for estimating the time of gap formation. Jpn J Ecol 31:307–316 (in Japanese with English synopsis)

Suzuki E (1981b) Regeneration of *Tsuga sieboldii* forest. IV. Temperate conifer forests of Kubotani-Yama and its adjacent area. Jpn J Ecol 31:421–434 (in Japanese with English Summary)

Tanaka N (1985) Patchy structure of a temperate mixed forest and topography in the Chichibu Mountains, Japan. Jpn J Ecol 35:153–167

Chapter 3
Pterocarya rhoifolia

Yosuke Nakano and Hitoshi Sakio

Abstract *Pterocarya rhoifolia* is a dominant canopy species of the Ooyamazawa riparian area of central Japan. In this study, to clarify how the life history of *P. rhoifolia* has adapted to riparian disturbance, we investigated the seed production, seedling regeneration, sprouting, distribution pattern, and size structure of this species in the Ooyamazawa riparian area. Annual seed production by mature *P. rhoifolia* fluctuated between years, with seeds produced during most years from 1995 to 2014. *P. rhoifolia* diameter at breast height (DBH) showed continuous distribution from saplings to mature trees, with some peaks. *P. rhoifolia* trees formed patches of various sizes along streams, with large, high-density patches observed on large landslide deposits; stands on such patches were generally even in age. Although current-year *P. rhoifolia* seedlings were found on litter, gravel, mineral soil, and fallen logs, almost all such seedlings died within several years. We found up to 10 sprouts per *P. rhoifolia* individual, including a few large sprouts per stem; however, no significant correlation was detected between sprout number and the DBH of the main stem. Therefore, we conclude that *P. rhoifolia* populations are generally maintained by seedling regeneration following large riparian disturbances in the Ooyamazawa riparian area.

Keywords Japanese wingnut · Life history · Seed production · Seedling regeneration · Spatial pattern · Disturbance regime

Y. Nakano (✉)
Tadami Beech Center, Fukushima, Japan
e-mail: nkn.yhsk@gmail.com

H. Sakio
Sado Island Center for Ecological Sustainability, Niigata University, Niigata, Japan
c-mail: sakio@agr.niigata-u.ac.jp; sakiohit@gmail.com

© The Author(s) 2020   39
H. Sakio (ed.), *Long-Term Ecosystem Changes in Riparian Forests*, Ecological Research Monographs, https://doi.org/10.1007/978-981-15-3009-8_3

3.1   Introduction

The genus *Pterocarya* belongs to the family Juglandaceae and was once widely distributed in the Northern Hemisphere, with fossils dating to the early Oligocene (Manchester 1987, 1989; Manos et al. 2007; Kozlowski et al. 2018). Currently, the extant *Pterocarya* comprise six generally accepted species, with one species in western Eurasia (Turkey, Georgia, Azerbaijan, and Iran) and five others in eastern Asia (China, the Korean Peninsula, Laos, Vietnam, Taiwan, and Japan) (Kozlowski et al. 2018). All of those *Pterocarya* species are elements of riparian forests in each region. In Japan, Juglandaceae comprises three species in three genera: *Pterocarya rhoifolia* Sieb. et Zucc., *Platycarya strobilacea* Sieb. et Zucc., and *Juglans mandshurica* Maxim. var. *sachalinensis* (Komatsu) Kitam.

The Japanese wingnut *P. rhoifolia* is the only volunteer species of *Pterocarya* in Japan. Although reports refer to the presence of an adjunct *P. rhoifolia* population from China (Laoshan, East Shandong), there is no reliable record and the presence of this species in China remains unclear (Kozlowski et al. 2018). The standard Japanese name for *P. rhoifolia* is "*Sawagurumi*," which is derived from the fact that this species grows near mountain streams. *P. rhoifolia* is a representative riparian element in the cool temperate forest of Japan, is widely distributed from southern Hokkaido to Kyusyu, and grows in regions with heavy and light snowfall (Hotta 1975; Kawahara et al. 2009; Nakano and Sakio 2017; Fig. 3.1). In snowy regions on the Japan Sea side and Tohoku region (northeastern Honshu), *P. rhoifolia* forms the riparian forest canopy with *Aesculus turbinata* Blume and *Cercidiphyllum japonicum* Sieb. et Zucc. (Kikuchi 1968; Suzuki et al. 2002). On the Pacific Ocean side, *Fraxinus platypoda* Oliv. coexists with these species (Sakio 1997; Sakio et al. 2002).

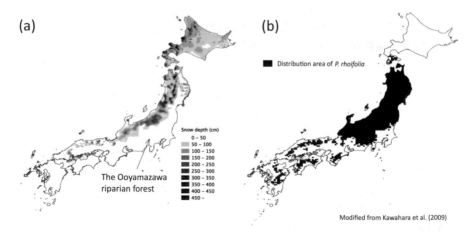

Fig. 3.1 (**a**) Annual maximum snow depth in Japan depth derived from Japanese Meteorological Agency mesh climate data (Nakano and Sakio 2017 revised), (**b**) map of the *P. rhoifolia* distribution

Pterocarya rhoifolia forests develop on stream banks, mud flow terraces, and flood terraces in floodplains, and on the alluvial cone and talus in lower hill slopes along mountain streams (Sato 1988; Kaneko and Kawano 2002). These habitats characteristically undergo land-surface disturbance, e.g., floods, debris flows, and landslides. Therefore, P. rhoifolia maintains its populations by adapting its life history to disturbance regimes in riparian areas.

In the Ooyamazawa riparian area on to the Pacific Ocean side of Japan, P. rhoifolia forms the canopy with F. platypoda and C. japonicum. Although P. rhoifolia dominates the canopy with F. platypoda in the Ooyamazawa riparian area, P. rhoifolia canopy trees also grow in patches along streams. In this chapter, we clarify the life history of P. rhoifolia in riparian forest based on studies in the Ooyamazawa riparian forest, and discuss the adaptation of the P. rhoifolia life history to riparian disturbances.

3.2 Study Species

Pterocarya rhoifolia is a deciduous canopy tree species with straight trunks and can reach 35 m in height and 1.2 m in diameter at breast height (DBH) (Hayashi 1969). It has a single large trunk or a few large trunks in a stool. It often produces sprouts, although not as frequently as C. japonicum. In the Ooyamazawa riparian forest, almost all P. rhoifolia had a single large trunk in a stool. In 2011, we recorded a maximum DBH and tree height of P. rhoifolia of 89.8 cm and 39.9 m, respectively (Fig. 3.2). Its bark is gray and split lengthwise in the mature tree stage. The terminal bud has 1–3 bud scales that fall between autumn and winter and overwinters as a naked bud with pubescence. Its leaves are alternate and imparipinnate compound, 20–30 cm long. The leaves have 9–21 leaflets (Fig. 3.3). The petiole and rachis are finely pubescent. The leaf rachis of the closely related species Pterocarya stenoptera is often winged, while that of P. rhoifolia is not.

3.3 Reproductive Characteristics

Pterocarya rhoifolia is a monoecious species with unisexual flowers. It is an anemophilous species, and flowers bloom and new leaves develop simultaneously. P. rhoifolia produces a female catkin hanging from the terminal part of a current-year shoot and some green male catkins hanging from the axil at the base of a current-year shoot (Fig. 3.4). It flowers in late May in the Ooyamazawa riparian forest. Although the male catkins fall off after pollen dispersal, the female catkins develop into infructescences. One infructescence has 20–60 fruit (Kaneko 2009; Figs. 3.5 and 3.6). One P. rhoifolia fruit is a nut with two wings that develop from bractlets; hence, the English name of P. rhoifolia is "Japanese wingnut." The nut is about 0.8 cm in size, and about 2.2 cm including the wings; the oven-dried nut and

Fig. 3.2 The typical mature tree form of *P. rhoifolia* (in the Ooyamazawa riparian forest)

nut including wings of *P. rhoifolia* (mean ± SD) weighed 70 ± 8 and 90 ± 11 mg, respectively (Sakio et al. 2002). A fruit contains one seed. In the Ooyamazawa riparian forest, *P. rhoifolia* fruit matures and is dispersed by wind from September to November (Fig. 3.7).

Pterocarya rhoifolia can live for up to 150 years (Kaneko 2009) and starts producing fruit at 40–80 years, at which point the DBH is 28 cm and the tree height is 16 m (The Japanese Riparian Forest Research Group 2001). Although *P. rhoifolia* reaching the canopy layer with stems larger than 30 cm DBH can bloom, small-diameter individuals released from suppression and understory individuals cannot bloom (Kaneko 2009). Therefore, to reach reproductive maturity, *P. rhoifolia* requires sufficient size and light.

Mature *P. rhoifolia* produce seeds most years, although the annual seed production fluctuates (Sawada et al. 1998; Kaneko and Kawano 2002; Sakio et al. 2002). In

Fig. 3.3   The leaves of *P. rhoifolia*

Fig. 3.4   A flowering branch of *P. rhoifolia* with male and female catkins

Fig. 3.5 Fruiting individual of *P. rhoifolia*

Fig. 3.6 A branch with leaves and a fruiting spike

the Ooyamazawa riparian forest, *P. rhoifolia* produced seeds almost every year between 1995 and 2014, although very few were produced in 1996 and 2006 (Fig. 3.8). Moreover, the annual variation in seed production has tended to alternate yearly since 2005. The coefficient of variation (CV) between 2005 and 2014 was

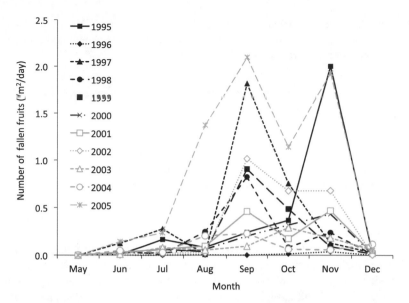

Fig. 3.7 Seasonal change in amount of fallen *P. rhoifolia* fruit in the Ooyamazawa riparian forest (1995–2005)

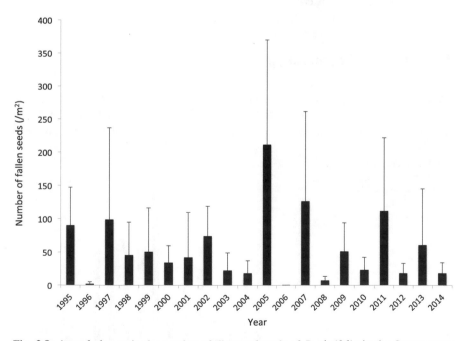

Fig. 3.8 Annual change in the number of dispersed seeds of *P. rhoifolia* in the Ooyamazawa riparian forest (1995–2014). Vertical bars indicate the standard deviation

larger than that between 1995 and 2004 ($CV_{1995-2004} = 0.75$, $CV_{2005-2014} = 0.89$). The variation in annual seed production in *P. rhoifolia* is intermediate between that of *F. platypoda* and *C. japonicum* (Sakio et al. 2002).

3.4 Structure and Distribution

According to Sakio et al. (2002), a 4.71 ha study plot in the Ooyamazawa riparian forest contained 112 *P. rhoifolia* over 4 cm in DBH, of which 80 (71.4%) were in the canopy layer. The mean DBH of the *P. rhoifolia* canopy trees was 44.6 ± 11.5 cm, with a maximum of 77.7. The DBH distribution of the *P. rhoifolia* population was continuous from saplings to large individuals (Fig. 3.9), but there were some peaks: one for small trees <5 cm in DBH (suggesting that *P. rhoifolia* maintains sapling banks) and another at 35–40 cm in DBH (suggesting that synchronous regeneration of *P. rhoifolia* has occurred several times in the past). The trees form patches of various sizes along a stream, with some large high-density patches on large landslide deposits (Sakio et al. 2002; Fig. 3.10). The mean age of the *P. rhoifolia* trees forming these patches is about 90 years and most are even-aged. These results suggest that *P. rhoifolia* invades the sites of large disturbances, such as floods, debris flows, and landslides, where it forms colonies.

3.5 Seedling Regeneration

The dispersed seeds of *P. rhoifolia* generally germinate in the spring of the following year. *P. rhoifolia* forms almost no soil seed bank. Kubo et al. (2008) investigated the species composition of the soil seed bank (to 5 cm depth) for the Ooyamazawa

Fig. 3.9 Frequency distribution of the DBH of *P. rhoifolia* in a 4.71-ha study plot in the Ooyamazawa riparian forest

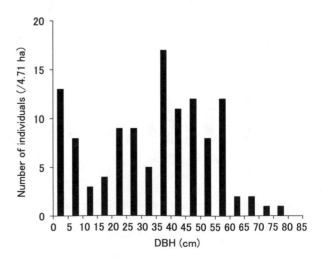

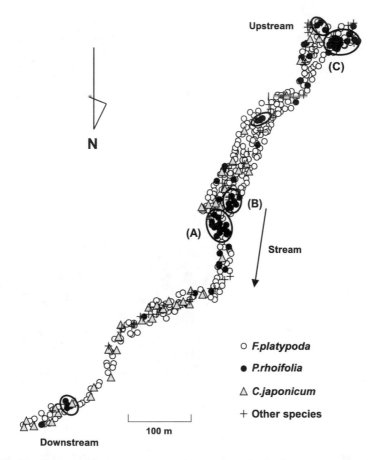

Fig. 3.10 Distribution of three canopy trees along a stream in the Ooyamazawa riparian forest. Patches (A–C) where a large landslide occurred about 90 years ago were dominated by *P. rhoifolia* canopy trees with a few *C. japonicum* trees (Sakio et al. 2002)

riparian forest and found only one buried viable *P. rhoifolia* seed in 30 L soil. The current-year seedlings of *P. rhoifolia* are epigeal cotyledons type, which have two opposite cotyledons deeply palmately four-clefted at germination (Fig. 3.11). Their alternate primary leaves are simple leaves until the first or second leaf stage, and become imparipinnate compound leaves at larger leaf stages.

The germination sites of current-year seedlings of *P. rhoifolia* in the Ooyamazawa riparian forest are on litter, gravel, mineral soil, and fallen logs (Sakio et al. 2002). On Chichibu Mountain, *P. rhoifolia* seeds germinate both under closed canopy and in gaps in the riparian forest dominated by *F. platypoda* and *P. rhoifolia* (Kisanuki et al. 1995). These results suggest that the germination sites of *P. rhoifolia* in riparian areas are not markedly limited.

In the Ooyamazawa riparian forest, current-year seedlings that germinate on litter tend to die in the current year because of low light intensity and drought, while

Fig. 3.11 Current-year
seedling of *P. rhoifolia* with
four-part cotyledons

current-year seedlings on gravel, mineral soil, and fallen logs die within 3 years
(Sakio et al. 2002). Furthermore, almost all current-year *P. rhoifolia* seedlings that
germinate under a closed canopy die in the current year (Kisanuki et al. 1995).
Taking this into consideration, under what conditions can current-year seedlings of
P. rhoifolia survive and grow? Nursery experiments have indicated that the height of
current-year *P. rhoifolia* seedlings increases with relative light intensity (RLI),
reaching about 30 cm at over 40% RLI (Sakio et al. 2008). An investigation of the
relationship between the distribution of *P. rhoifolia* saplings (height >1 m and DBH
<4 cm) and topography in the Ooyamazawa riparian forest found that more saplings
were distributed in abandoned channels and floodplains than on hillslopes or in
active channels (Sakio et al. 2002, Table 3.1; Fig. 3.12). The abandoned channels
and floodplains were directly affected by stream flow and lacked an understory.
Moreover, the microhabitats in the Ooyamazawa riparian forest in which *P. rhoifolia*
saplings (≥1 years old) tend to thrive consist of deposited gravel with thin soil and
no herbaceous layer (Kubo et al. 2000).

There are no statistical relationships between canopy gaps and the distribution of
P. rhoifolia saplings (height >1 m and DBH <4 cm) in the Ooyamazawa riparian
forest (Table 3.1; Fig. 3.13). However, Sato (1992) investigated the regeneration of
P. rhoifolia saplings in a *P. rhoifolia* forest in Hokkaido and found a weak positive
correlation between the distribution of *P. rhoifolia* saplings (2 m ≤ tree height ≤ 4 m)

Table 3.1 Number of grids with *P. rhoifolia* saplings (height ≥ 1 m) and the mean number of saplings per grid (Sakio et al. 2002 revised)

Grid condition	Presence grids of saplings (%)	Total number of saplings	Mean number of saplings per grid ($\pm$s.d.)
Topography (number of grids)			ns[a]
Hillslope (428)	13 (3.3%)	17	1.13 $\pm$ 0.35
Active channel (161)	9 (5.6%)	10	1.11 $\pm$ 0.33
Abandoned channel (208)	44 (21.2%)	71	1.61 $\pm$ 0.97
Floodplain (553)	55 (9.9%)	94	1.71 $\pm$ 1.54
Canopy (number of grids)			ns[b]
Under canopy (1052)	96 (9.1%)	157	1.64 $\pm$ 1.30
Gap (298)	27 (9.1%)	35	1.30 $\pm$ 0.72

s.d. standard deviation, *ns* not significant
[a]Results of Tukey–Kramer test
[b]Results of *t*-test

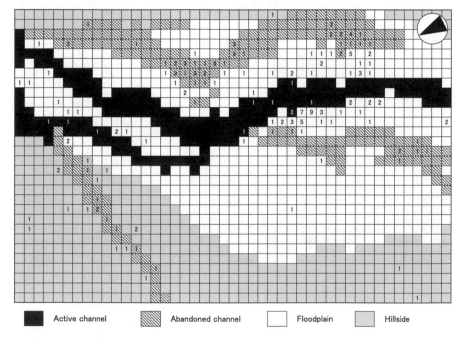

Active channel   Abandoned channel   Floodplain   Hillside

Fig. 3.12 Microtopography and distribution of *P. rhoifolia* saplings in a 0.54-ha study plot in the Ooyamazawa riparian forest (Sakio et al. 2002 revised). This plot was divided into 1350 quadrats (2×2 m^2). The microtopography was recorded in each quadrat and was classified into the following categories: hillside, active channel, abandoned channel, and floodplain. The figures show the numbers of *P. rhoifolia* saplings in each quadrat

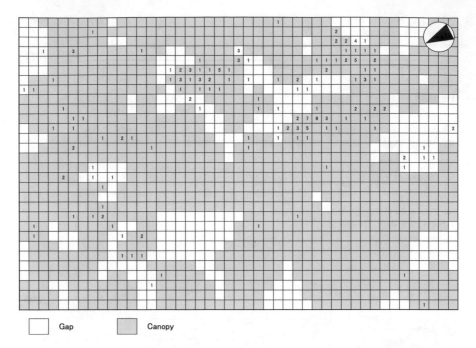

Gap   Canopy

Fig. 3.13 Canopy gaps and distribution of *P. rhoifolia* saplings in a 0.54-ha study plot in the Ooyamazawa riparian forest. This plot was divided into 1350 quadrats (2×2 m^2)

and relative illumination, and reported that *P. rhoifolia* saplings were distributed contiguously around canopy gaps. Moreover, Sakio (1993) investigated the pattern of leaf expansion and shoot elongation of *P. rhoifolia* saplings in Ooyamazawa riparian forest, and found that *P. rhoifolia* saplings (height = 80.3 ± 13.3 cm) in a canopy gap continued to expand their leaves and develop current-year shoots from May to August.

The factors that enable the survival and growth of *P. rhoifolia* seedlings and saplings have not been explained completely. However, many investigations have suggested that, for *P. rhoifolia* to regenerate seedlings, bare ground caused by land-surface disturbance with destruction of the herbaceous layer and a canopy gap are probably needed.

3.6 Sprouting Traits

Pterocarya rhoifolia produces some sprouts from the root collar and lowest part of the stem, but lacks root suckers. In the Ooyamazawa riparian forest, *P. rhoifolia* has a maximum of 10 sprouts/individual and there are no significant correlations between the number of sprouts and DBH of the main stem (Sakio et al. 2002). In an unpublished study, overall, 79.2% of individuals had sprouts, 95.3% of which

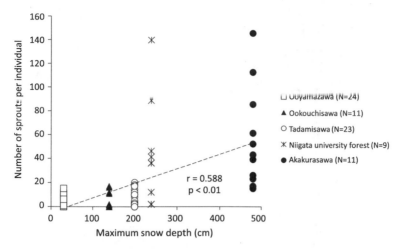

Fig. 3.14 Relationship between the maximum snow depth and number of sprouts per mature *P. rhoifolia*. *N* number of individuals (Nakano and Sakio 2017 revised)

were less than 50 cm high (Y. Nakano, unpublished data). In the Ooyamazawa riparian forest, such sprouts have no role in maintaining the population (Nakano and Sakio 2017).

However, the sprouting traits of *P. rhoifolia* change in response to maximum snow depth. Nakano and Sakio (2017) comparing several areas with maximum snow depths from 30 to 480 cm in a cool temperate mountainous area in central Japan, which includes the Ooyamazawa riparian forest as a low-snow region, and reported that the number of sprouts per *P. rhoifolia* individual increased with maximum snow depth (Fig. 3.14). In deep snow, the sprouts of *P. rhoifolia* may play a role in repairing individuals damaged by the snowpack to maintain the population (Nakano and Sakio 2018). They also found that, with increasing maximum snow depth, the DBH decreased, maximum stem length and tree height shortened, trees tended toward a "dwarf shrub" form, and seed production decreased (Nakano and Sakio 2017). These results suggest trade-offs between clonal growth (producing sprouts) and sexual reproduction (seed production) and between producing sprouts and height growth (Nakano and Sakio 2017; Fig. 3.15). This sprouting ability could conceivably be due to *P. rhoifolia* growing in an environment that tends to be influenced by disturbances specific to riparian areas, such as floods and landslides. Therefore, the sprouting ability of *P. rhoifolia* is insurance enabling it to survive when threatened.

Ito (1992) investigated dry matter partitioning in seedlings and sprouts of *P. rhoifolia* and showed that the slopes of the regression lines of the allometric relationship between basal trunk diameter and tree height were greater in sprouts than in seedlings, suggesting a change in growth characteristics from the "waiting" type (diameter-preferred growth) in seedlings to the "competing" type (elongation-preferred growth) in sprouts.

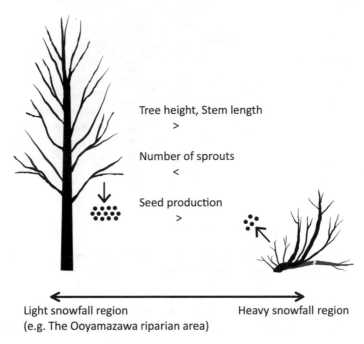

Tree height, Stem length

>

Number of sprouts

<

Seed production

>

Light snowfall region   Heavy snowfall region
(e.g. The Ooyamazawa riparian area)

Fig. 3.15 Schematic diagram relating the life history traits of *P. rhoifolia* and snowfall conditions (Nakano and Sakio 2017 revised)

3.7 Conclusion

In the Ooyamazawa riparian forest, *P. rhoifolia* forms the canopy with *F. platypoda* and *C. japonicum*. Sato (1992) compared the distribution, growth, and survival of saplings of the main canopy tree species (*P. rhoifolia, Acer mono, Alnus hirsute*, and *Ulmus laciniata*) in a *P. rhoifolia* forest in Hokkaido. Defining a pioneer species by its high germinating capacity, high growth rate, and high survival rate under sufficient light, they concluded that *P. rhoifolia* is not a typical pioneer species, but is intermediate between a pioneer and non-pioneer species. In the Ooyamazawa riparian forest, *P. rhoifolia* is more of a pioneer than *F. platypoda* and *C. japonicum*, particularly in terms of the growth rate during the seedling and sapling stages (cf. Chap. 7).

Mature *P. rhoifolia* produce and disperse seeds almost every year. *P. rhoifolia* seeds invade the sites of large disturbances, including floods, debris flows, land-slides, and canopy gaps, and establish there. Moreover, *P. rhoifolia* can grow rapidly with sufficient light. Consequently, this tree species can form the canopy and even-aged populations in the Ooyamazawa riparian forest. In this forest, *P. rhoifolia* does not produce many sprouts. Hence, sprouts may not contribute to maintaining *P. rhoifolia* individuals for a long time, such as *C. japonicum* (cf. Chap. 4), and the lifespan of *P. rhoifolia* appears to be about 150 years (Kaneko 2009). Consequently, the dominance ratio of late-successional tree species in the canopy such as *F. platypoda* and *C. japonicum* may increase after the lifespan of *P. rhoifolia* without large disturbances. However, if large disturbances occur along rivers and generate suitable sites for *P. rhoifolia* regeneration, even-aged *P. rhoifolia* forest may reestablish (Sato 1988). Therefore, the life history traits of *P. rhoifolia* are adapted to disturbances in riparian areas and the species may be able to maintain its population in the Ooyamazawa riparian forest.

References

Hayashi Y (1969) Atlas of useful tree (forest tree part). Seibundoushinkousya, Tokyo, 472 pp (in Japanese)

Hotta M (1975) Tree species in hills and fields (II). Hoikusya, Tokyo (in Japanese)

Ito S (1992) Dry matter partitioning in seedlings and sprouts of several deciduous broad-leaved tree species with special reference to the functions of tree architecture. Bull Kyushu Univ For 66:19–30

Kaneko Y (2009) *Pterocarya rhoifolia* Sieb.et Zucc. (Juglandaceae). In: Japan Botanical History Editing Committee (ed) Botanical history of trees in Japan. Japan Forestry Investigation Committee, Tokyo, pp 353–386 (in Japanese)

Kaneko Y, Kawano S (2002) Demography and matrix analysis on a natural *Pterocarya rhoifolia* population developed along a mountain stream. J Plant Res 115:341–354

Kawahara T, Watanabe T, Matsui T, Takahashi M (2009) Distribution map of *Pterocarya rhoifolia*. In: Japan Botanical History Editing Committee (ed) Botanical history of trees in Japan. Japan Forestry Investigation Committee, Tokyo, p 746 (in Japanese)

Kikuchi T (1968) Forest communities along the Oirase valley, Aomori prefecture. Ecol Rev 87:87–94

Kisanuki H, Kaji M, Suzuki K (1995) The survival process of ash (*Fraxinus spaethiana* Ling.) and wingnut (*Pterocarya rhoifolia* Sieb. et Zucc.) seedlings at the riparian forest at Chichibu mountains. Bull Univ Tokyo For 93:49–57 (in Japanese with English summary)

Kozlowski G, Bétrisey S, Song Y (2018) Wingnuts (*Pterocarya*) & walnut family. Natural History Museum Fribourg, Switzerland

Kubo M, Shimano K, Sakio H, Ohno K (2000) Germination sites and establishment conditions of *Cercidiphyllum japonicum* seedlings in the riparian forest. J Jpn For Soc 82:349–354 (in Japanese with English summary)

Kubo M, Kawanishi M, Shimano K, Sakio H, Ohno K (2008) The species composition of soil seed banks in the Ooyamazawa riparian forest, in the Chichibu mountains, central Japan. J Jpn For Soc 90:121–124

Manchester SR (1987) The fossil history of the Juglandaceae. Monogr Syst Bot Missouri Bot Gard 21:1–137

Manchester SR (1989) Early history of the Juglandaceae. Plant Syst Evol 162:231–250

Manos PS, Soltis PS, Soltis DE, Manchester SR, Oh SH, Bell CD, Dilcher DL, Stone DE (2007) Phylogeny of extant and fossil Juglandaceae inferred from the integration of molecular and morphological data sets. Syst Biol 56:412–430

Nakano Y, Sakio H (2017) Adaptive plasticity in the life history strategy of a canopy tree species, *Pterocarya rhoifolia*, along a gradient of maximum snow depth. Plant Ecol 218:395–406

Nakano Y, Sakio H (2018) The regeneration mechanism of *Pterocarya rhoifolia* population in the heavy snowfall region of Japan. Plant Ecol 219:1387–1398

Sakio H (1993) Sapling growth pattern in *Fraxinus platypoda* and *Pterocarya rhoifolia*. Jpn J Ecol 43:163–167 (in Japanese with English summary)

Sakio H (1997) Effects of natural disturbance on the regeneration of riparian forests in a Chichibu Mountains, central Japan. Plant Ecol 132:181–195

Sakio H, Kubo M, Shimano K, Ohno K (2002) Coexistence of three canopy tree species in a riparian forest in the Chichibu mountains, central Japan. Folia Geobotanica 37:45–61

Sakio H, Kubo M, Shimano K, Ohno K (2008) Coexistence mechanisms of three riparian species in the upper basin with respect to their life histories, ecophysiology, and disturbance regimes. In: Sakio H, Tamura T (eds) Ecology of riparian forests in Japan: disturbance, life history and regeneration. Springer, Tokyo, pp 75–90

Sato H (1988) The structure and habitats of *Pterocarya rhoifolia* forest in Matsumae Peninsula, Southern Hokkaido, Japan. Jpn Soc For Envir 30:1–9 (in Japanese with English summary)

Sato H (1992) Regeneration traits of saplings of some species composing *Pterocarya rhoifolia* forest. Jpn J Ecol 42:203–214 (in Japanese with English summary)

Sawada H, Igarashi Y, Ohmura K, Kisanuki H, Kaji M (1998) Amounts of fallen samaras of *Fraxinus spaethiana* and *Pterocarya rhoifolia* at a riparian forest in the Chichibu Mountains. Trans Jpn For Soc 109:331–334 (in Japanese)

Suzuki W, Osumi K, Masaki T, Takahashi K, Daimaru H, Hoshizaki K (2002) Disturbance regimes and community structures of a riparian and an adjacent terrace stand in the Kanumazawa Riparian Research Forest, northern Japan. Forest Ecol Manag 157:285–301

The Japanese Riparian Forest Research Group (2001) Guideline for the management of riparian forests. Japan Forestry Investigation Committee, Tokyo (in Japanese)

Chapter 4
Cercidiphyllum japonicum

Masako Kubo and Hitoshi Sakio

Abstract The tertiary relict *Cercidiphyllum japonicum* is an important canopy tree species of riparian forests in Japan, despite typically occurring at low densities. Once mature, canopy individuals are typically 30 cm in diameter at breast height and have high annual seed production. Seedlings tend to germinate on steep slopes with exposed soil, but not in thick litter or gravel substrates. Annual seedling mortality is high at germination sites that are exposed to heavy rain and flooding. In contrast, *C. japonicum* has highly successful vegetative reproduction through basal sprouting (a product of endogenous [i.e., aging] and exogenous [i.e., gap formation and physical damage] factors), which leads to a multi-stemmed growth form. Sprouting allows *C. japonicum* to dominate stands over time, as it is longer lived than other coexisting species; this promotes the maintenance of its populations.

Keywords *Cercidiphyllum japonicum* · Life history · Riparian forest · Seedling traits · Seed production · Sprouting traits · Tertiary relict species

4.1 Introduction

Cercidiphyllum are known as tertiary relict species; fossils indicate that this angiosperm lineage arose during the Cretaceous period (Dosmann 1999). The genus comprises two genetically distinct dioecious tree species, *Cercidiphyllum japonicum* Siebold et Zucc. ex Hoffm. et Schult. and *Cercidiphyllum magnificum* (Nakai) Nakai (Fig. 4.1) (Li et al. 2002; Isagi et al. 2005). Although *Cercidiphyllum* was once distributed throughout the Northern Hemisphere (Crane and Stockey 1985;

M. Kubo (✉)
Faculty of Life and Environmental Science, Shimane University, Shimane, Japan
e-mail: kubom@life.shimane-u.ac.jp

H. Sakio
Sado Island Center for Ecological Sustainability, Niigata University, Niigata, Japan
e-mail: sakio@agr.niigata-u.ac.jp; sakiohit@gmail.com

© The Author(s) 2020   55
H. Sakio (ed.), *Long-Term Ecosystem Changes in Riparian Forests*, Ecological
Research Monographs, https://doi.org/10.1007/978-981-15-3009-8_4

Fig. 4.1 *Cercidiphyllum japonicum* and *Cercidiphyllum magnificum* at Ooyamazawa. (**a**) (photo by Kyoko Kato) and (**b**) show tree form; (**c**) and (**d**) show the leaves in spring and summer of *C. japonicum* and *C. magnificum*, respectively. The difference in leaf size between *C. japonicum* (right) and *C. magnificum* (left) is shown (**e**)

Skawińska 1986), *C. japonicum* is now found in only Japan and part of China, and *C. magnificum* is restricted to parts of Japan (Manchester et al. 2009). In Japan, *C. japonicum* mainly occurs in riparian areas in cool temperate forests on Hokkaido, Honshu, Shikoku, and Kyushu, and *C. magnificum* is found in subalpine forests on

Fig. 4.2 Multi-stemmed trunk and canopy of *Cercidiphyllum japonicum* in Hirogawara, Yamanashi

Honshu. There are no reports of varied morphology of *C. japonicum*, but the genotype is reported to vary between north-central and southwestern Japan (Qi et al. 2012). Populations of *C. japonicum* from north-central Japan have similar chloroplast DNA to *C. magnificum*, implying that some *C. japonicum* individuals may have had an advantage in severe climate conditions at northern latitudes, which could have facilitated the establishment and maintenance of northern populations (Qi et al. 2012).

Cercidiphyllum japonicum is an essential canopy species in Japanese riparian forests (Ohno 2008), although it is mostly found in low abundance and pure stands are rare. Saplings are not often observed, implying that germination may be limited in forests. The species is long lived, often forming large canopies and multi-stemmed trunks (stools) through sprouting (i.e., producing shoots from the base or from roots) (Fig. 4.2). It is possible that vegetative reproduction may compensate for low sapling recruitment, allowing *C. japonicum* to coexist at low density with other canopy species.

This species, called "*Katsura*" in Japanese, is a sacred tree with ancient associations to moon, mountain, and water deities. Therefore, many large, multi-stemmed individuals are designated as natural monuments throughout the country (Fig. 4.3). Although *C. japonicum* is an important species both culturally and ecologically, its life history remains poorly understood. It is thought that populations are maintained

Fig. 4.3 *Cercidiphyllum japonicum* as a natural monument in Japan. (**a**) Katsura of Takezaki in Shimane Prefecture; (**b**) Oo-Katsura of Itoi in Hyogo Prefecture; (**c**) Katsura of Idoi Shrine in Shiga Prefecture; (**d**) Senbon-Katsura of Inari Shrine in Iwate Prefecture; (**e**) Senbon-Katsura of Oohora in Iwate Prefecture. The Japanese common name of *C. japonicum* is "*Katsura*"

on microtopography features created by various disturbances in riparian forests over long periods. However, such old-growth forests are now scarce due to extensive logging, and understanding the life history of *C. japonicum*, including germination traits and survivorship, may be confounded by this habitat loss.

An undisturbed population of *C. japonicum* occurs in a 1-km long preserved forest in the Ooyamazawa riparian area in central Japan. Here, *C. japonicum* is found as a canopy species, with most individuals distributed on steep slopes and exposed rocks. This population is small relative to those of other canopy species such as *Fraxinus platypoda* (Chap. 2) and *Pterocarya rhoifolia* (Chap. 3). In this chapter, we clarify the life history of *C. japonicum* using observations from the Ooyamazawa population. We investigated distribution patterns, size structure, flowering and seed production, seedling regeneration traits, and sprouting traits, and considered life history strategies in adapting to riparian disturbances. Additionally, we partially compared the life history of *C. japonicum* to that of the closely related *C. magnificum*.

4.2 Study Species

Cercidiphyllum japonicum is a tall, straight-stemmed species up to 30 m in height and 1 m in diameter (Fig. 4.1a). Associated canopy species within riparian forests include *F. platypoda*, *P. rhoifolia*, *Aesculus turbinata*, and *Acer* spp. Many individuals have multi-stemmed trunks, with wide, shoot-producing rootstocks (stools) and expansive individual canopies (Fig. 4.2). Leaves are borne on short and long shoots and are atypical in shape; those produced in spring are heart shaped, and those produced in summer are rhomboid (Fig. 4.1c).

Cercidiphyllum magnificum is distributed with *Betula ermanii*, *Acer shirasawanum*, and *Pterostyrax hispida* in riparian areas of subalpine forests in north-central Japan, where various disturbances occur, including spring avalanches. Its stems are usually creeping, growing on average to 5 m in height and 10 cm in diameter, although large specimens can reach 20 m in height and 40 cm in diameter (Fig. 4.1b). Their leaves are larger than those of *C. japonicum* (Fig. 4.1e).

4.3 Structure and Distribution

Our inventory recorded 2111 individual trees within a 4.71-ha study plot in Ooyamazawa riparian area. Of these, 59 were *C. japonicum* individuals (12.6 individuals/ha) (Table 4.1) with a main stem ≥4 cm diameter at breast height (DBH), measured from the main (largest) stem in multi-stemmed individuals. Most individuals were in the canopy layer ($n = 47$), with nine in the subcanopy layer, and three in the shrub layer. The DBH of main stems ranged from 5 to 153 cm (Fig. 4.4).

Most individuals were distributed within a V-shaped valley in the study area (Fig. 4.5), which had 30-degree slopes on the valley sides and roughly 12-degree slopes in sedimentary debris flow areas (Chap. 1). Alluvial fan and terrace debris flows in upstream areas contained rich soil with a litter layer, but there was little litter on terrace scarps, new landslide sites, old landslide slopes, and talus in downstream areas (Kawanishi et al. 2004). Various disturbances, including erosion and

Table 4.1 *Cercidiphyllum japonicum* trees in Ooyamazawa with a main stem DBH of ≥4 cm

	Number of trees			Mean number of sprouts
	Canopy	Subcanopy	Shrub	
Female	20	0	0	8.9 ± 9.5
Male	26	0	0	12.5 ± 15.5
Immature	1	9	3	2.9 ± 4.4
Total	47	9	3	9.1 ± 12.2

Canopy-layer trees reach the canopy and are >20 m in height and 20 cm in DBH. Subcanopy layer trees do not reach the canopy and are <20 m high. Shrub layer trees are <10 m high. The mean number of sprouts reflects the mean ± standard deviation per individual

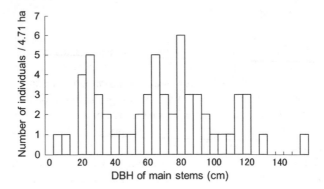

Fig. 4.4 Diameter at breast height size class distribution of main stems of *Cercidiphyllum japonicum* in Ooyamazawa. Individuals with a main stem DBH ≥4 cm are shown

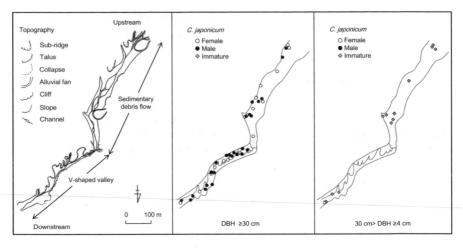

Fig. 4.5 Diagram of microtopography types and the distribution of *Cercidiphyllum japonicum* in Ooyamazawa. Individuals with a main stem DBH of ≥4 cm are shown

sedimentation of soil, sand, and/or gravel, were frequent in downstream valley areas due to stream flow and steep slopes. Upstream upland areas were less disturbed, whereas the valley bottom was filled in by previous large landslides and/or debris flows. Both the tree density and total basal area of *C. japonicum* were greater in the V-shaped valley than in the sedimentary debris flow areas upstream, and individuals were distributed over microtopography features such as sub-ridges, talus, and collapses (Kubo et al. 2001a). Only 46 of the 59 individuals, 20 females and 26 males (Table 4.1), were mature, and female and male trees were randomly distributed relative to each other (Fig. 4.5).

Individuals were distributed along a stream in the study area from 1200 to 1600 m in elevation, and were found co-occurring with *C. magnificum* on a talus slope next to a stream above 1600 m (Fig. 4.6). *Cercidiphyllum japonicum* was found at <1650 m in elevation, but *C. magnificum* was found at elevations ≥1600 m, extending to the ridge border at approximately 1720 m in elevation. The co-occurring

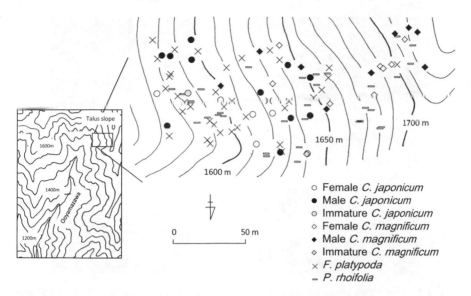

Female *C. japonicum*
Male *C. japonicum*
Immature *C. japonicum*
Female *C. magnificum*
Male *C. magnificum*
Immature *C. magnificum*
F. platypoda
P. rhoifolia

Fig. 4.6 Distribution of *Cercidiphyllum japonicum* and *Cercidiphyllum magnificum* on an upper stream talus slope at Ooyamazawa (Kubo et al. 2010, revised). Individuals with a main stem DBH of ≥4 cm are shown

canopy species *F. platypoda* and *P. rhoifolia* were distributed throughout the elevation range, with a greater number of *P. rhoifolia* individuals on the upper slopes of the study area.

4.4 Reproductive Traits

4.4.1 Flower

The canopy of *C. japonicum* turns purple-red at the time of blooming (Fig. 4.7), when individuals produce a large number of female or male flowers that lack a perianth at the tips of short shoots (Fig. 4.8). Blooming occurs in late April or early May in Ooyamazawa.

We categorized the 59 individuals of *C. japonicum* according to the DBH of their main stem (Fig. 4.9). All female and male trees reached the canopy layer, and except for one individual, all immature trees were found in the subcanopy and shrub layers (Table 4.1). All immature trees were < 26 cm in DBH, but one immature tree with a DBH of only 21 cm reached the canopy layer (Kubo and Sakio, unpublished data), implying that stem size may relate to reproductive maturity. Observations of the 46 reproductive individuals from 2000 to 2007 indicated that all mature trees with large main stems or many-stemmed trunks, including very old specimens, flowered and fruited heavily every year (Kubo and Sakio, unpublished data).

Fig. 4.7 Female *Cercidiphyllum japonicum* with purple-red flowers in late March in Shimane Prefecture (photo by Takuya Kashima)

The age of reproductive maturity varies widely among tree species. Many pioneer species begin flowering before 10 years, whereas late successional species such as beech and oak begin around age 60 (Thomas 2000). *Fraxinus platypoda* reaches reproductive maturity around age 50, and *P. rhoifolia* at age 20 (Sakio, personal observation). Although we cannot precisely determine the age of reproductive maturity for *C. japonicum*, four immature individuals in the subcanopy layer ranged from 16.9 to 22.0 cm in DBH and from 86 to 88 years in age (Sakio et al. 2002). By estimation, it may take up to 100 years for *C. japonicum* to reach a DBH of 30 cm.

4.4.2   Seed Production

Leaf-out occurs earlier in *C. japonicum* than in *F. platypoda* and *P. rhoifolia*, beginning in early May. Leaves appear soon after the very short blooming period, which generally lasts 10 days. Following flowering, banana-shaped follicles (fruits) are produced, each with approximately 20 samaras (seeds). These seeds are small, approximately 6 mm in length, including the wing, and 2 mm in width (Fig. 4.10). The dry weight of individual seeds collected from Ooyamazawa was 0.58 ± 0.14 mg ($n = 20$; Sakio et al. 2002).

Between late October and early November, *C. japonicum* leaves turn yellow (Fig. 4.11). This is followed by seasonal defoliation and seed dispersal (Fig. 4.12). Seed dispersal may begin as early as July, but it has been suggested that most seeds are immature at that time (Mizui 1993). Seeds are wind-dispersed, with larger dispersal events observed after seasonal defoliation. In the study area, the number

Fig. 4.8   Female and male flowers of *Cercidiphyllum japonicum* (photo by Yasuo Iizuka)

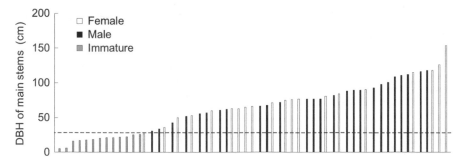

Fig. 4.9   Female, male, and immature *Cercidiphyllum japonicum* main stem DBH sizes at Ooyamazawa. The dotted line indicates a 26-cm DBH. Individuals with a main stem DBH of ≥4 cm are shown

Fig. 4.10 Fruit and seed of
Cercidiphyllum japonicum

of seeds was greatest from late October to mid-November and the dry weight of fallen leaves was greatest in October. Although seed production varied among years, *C. japonicum* produced seeds annually without an observed poor crop year (Fig. 4.13).

It is probable that the age of reproductive maturity of *C. japonicum* is later than those of other tree species; however, *C. japonicum* produces seeds annually after reaching maturity. Reproductive maturation at roughly 100 years old may be optimal in the context of the long lifespan of *C. japonicum*, as mature individuals can produce high annual volumes of seed over a long period.

4.5   Seedling Regeneration

4.5.1   Seedling Emergence

Although this species produces a large number of seeds annually, saplings are seldom observed in the understory of closed forests. Therefore, we wondered

Fig. 4.11 *Cercidiphyllum japonicum* leaves turning yellow in October at Ooyamazawa. (**a**) and (**b**) *C. japonicum*; (**c**) *C. magnificum*

when and where seedlings emerge, and investigated microhabitat conditions for *C. japonicum* seedlings. Numerous seedlings were recorded on the forest floor in

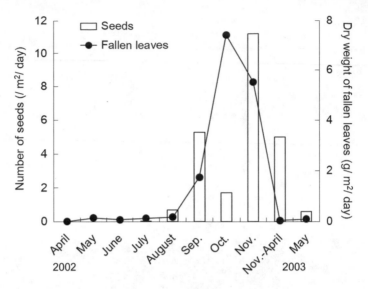

Fig. 4.12 Seasonal change in the amount of fallen leaves and dispersed seeds of *Cercidiphyllum japonicum*

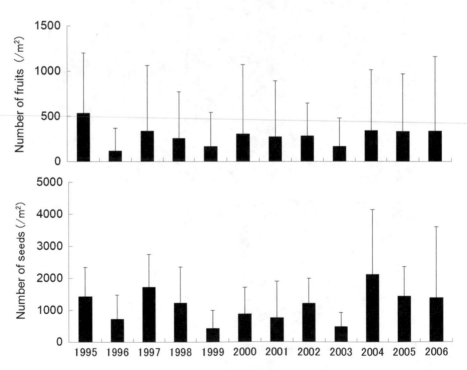

Fig. 4.13 Annual change in the number of dispersed seeds and fruits of *Cercidiphyllum japonicum*. Vertical bars indicate the standard deviation

Fig. 4.14 Seedling germination sites of *Cercidiphyllum japonicum*. (**a**) bare soil; (**b**) bare soil with small gravel; (**c**) a fallen log covered with moss; (**d**) a fallen log; (**e**) rock; and (**f**) first-year seedling size. Arrows and circles highlight *C. japonicum* seedlings and saplings

late May, after canopy trees had leafed out. Seedlings emerged on bare soil, rocky ground, and fallen logs (Fig. 4.14), but not on litter or gravel (Kubo et al. 2000). Such germination sites, where litter cannot accumulate and bare soil is exposed, occurred on steep slopes at small scales (Kubo et al. 2000), and saplings were frequently observed on rocks (Fig. 4.14e).

These observations are consistent with the results of a nursery experiment (Kubo et al. 2004). We investigated *C. japonicum* emergence and seedling survival in a nursery for 21 months, in which treatments of bare soil, soil with litter, and gravel under 3.0, 10.9, 22.7, 60.1, and 100.0% relative photosynthetic photon flux density (RPPFD) light conditions were tested. Seedling emergence was greatest in the bare-soil treatment (Fig. 4.15).

Some seeds at Ooyamazawa landed in areas appropriate for germination, i.e., bare soil on steep slopes. However, those which fell in areas with high litter or gravel accumulation were unlikely to survive after germination given the small size of the seedlings (approximately 1.5 cm high; Fig. 4.14f). Generally, seedlings from small seeds cannot penetrate a thick litter layer (Seiwa and Kikuzawa 1996). Furthermore, severe evaporation is more pronounced in gravel than in soil (Kayane 1980). In the nursery, seedlings preferentially emerged from gravel under the two lowest light conditions (3.0 and 10.9% RPPFD; Fig. 4.15), potentially because they were able to maintain moisture levels under those light conditions.

It is possible that *C. japonicum* has some form of seed dormancy (Kubo et al. 2008). We conducted germination tests using topsoil across three microhabitat conditions at Ooyamazawa, and although germination was observed, successful recruitment was very low (Table 4.2). Furthermore, under nursery conditions, four seedlings emerged in the second year after seeding (three seedlings and one seedling under 10.9 and 22.7% RPPFD, respectively, in bare soil; Fig. 4.15).

4.5.2 Seedling Survival

A good place for seedling emergence may not be a good place for seedling survival (Farmer 1997); forest conditions may be too severe for *C. japonicum* seedlings to survive. Although seedlings emerged under low light conditions (less than 10% relative illuminance), almost all were dead by summer (Fig. 4.16). The first-year survival rate of *C. japonicum* seedlings appeared to be greater under higher light conditions, although many individuals died by autumn. Seedling height and leaf size were reduced under lower light conditions (Kubo et al. 2000), and it is probable that smaller seedlings are more likely to die, be washed away, or to be covered by litter.

Cercidiphyllum japonicum seedlings emerged in bare soil under all light conditions in the nursery experiment, but seedling survival was greatest under moderate (10.9% RPPFD) and high light conditions (treatments of 22.7 and 60.1% RPPFD; Fig. 4.15). Extreme light conditions (3.0 and 100.0% RPPFD) were not optimal to survival. First-year seedlings grew poorly under high light (60.1% RPPFD), although most survived. After the second year, the seedlings appeared able to tolerate high light and grew tallest under this treatment (60.1% RPPFD; Fig. 4.17).

Slope may play a role in the observed differences in survival between nursery and forest conditions. The seedbeds in the nursery were flat, promoting seedling survival under moderate light conditions. Although *C. japonicum* seedlings likely have some shade tolerance, larger seedlings growing under bright light conditions would have a

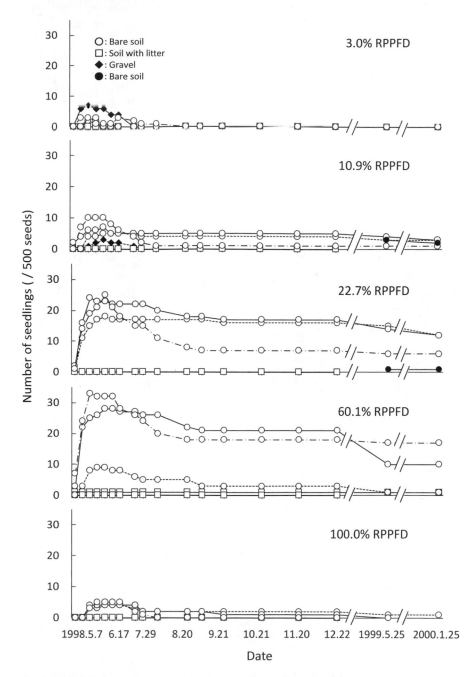

Fig. 4.15 Seedling germination and survival under experimental soil and light treatments (Kubo et al. 2004, revised). Open circles show seedlings grown on bare soil in 1998, open squares show seedlings grown on soil with litter in 1998, black rhomboids show seedlings grown on gravel in 1998, and black circles show new seedlings grown on bare soil in 1999

Table 4.2 Composition of seedlings emerging from topsoil and/or litter layers in Ooyamazawa (Kubo et al. 2008, revised)

	Topsoil				Litter layer			
	Upstream terrace of debris flow	Downstream terrace of debris flow	Downstream lower side slope		Upstream terrace of debris flow	Downstream terrace of debris flow	Downstream lower side slope	
Woody plant								
Hydrangea macrophylla var. acuminata	30.2 ± 35.2 (151)	0.2 ± 0.4 (1)	5.4 ± 3.6 (27)	*	3.6 ± 3.4 (18)	0.0 ± 0.0	1.0 ± 0.8 (5)	*
Buddleja japonica	0.8 ± 1.1[b] (4)	8.2 ± 5.9[a] (41)	0.0 ± 0.0[b]		0.2 ± 0.3 (1)	0.6 ± 0.7 (3)	0.0 ± 0.0	
Euptelea polyandra	0.0 ± 0.0[b] (1)	4.8 ± 1.8[a] (24)	3.0 ± 1.9[a] (15)	*	0.4 ± 0.4 (2)	2.0 ± 1.7 (10)	1.8 ± 1.3 (9)	*
Betula grossa	5.8 ± 3.6[a] (29)	0.6 ± 0.9[b] (3)	1.2 ± 1.1[b] (6)	*	4.2 ± 1.4[a] (21)	1.4 ± 0.8[b] (7)	0.8 ± 0.5[b] (4)	*
Actinidia arguta	2.2 ± 2.3 (11)	0.4 ± 0.9 (2)	0.4 ± 0.5 (2)	*	2.6 ± 1.6 (13)	2.2 ± 1.9 (11)	0.0 ± 0.0	*
Cercidiphyllum japonicum	0.6 ± 0.9 (3)	1.0 ± 1.2 (5)	0.8 ± 0.4 (4)	*	4.6 ± 4.5 (23)	5.4 ± 3.0 (27)	1.0 ± 0.7 (5)	*
Rubus phoenicolasius	0.0 ± 0.0[b]	2.2 ± 1.8[a] (11)	0.2 ± 0.4[b] (1)		0.0 ± 0.0	0.6 ± 0.7 (3)	0.2 ± 0.3 (1)	
Pterocarya rhoifolia	0.2 ± 0.4 (1)	0.0 ± 0.0	0.0 ± 0.0	*	0.0 ± 0.0	0.4 ± 0.4 (2)	0.0 ± 0.0	*
Schizophragma hydrangeoides	0.0 ± 0.0	0.2 ± 0.4 (1)	0.0 ± 0.0	*	0.0 ± 0.0	0.2 ± 0.3 (1)	0.0 ± 0.0	*
Pterostyrax hispidus	0.2 ± 0.4 (1)	0.0 ± 0.0	0.0 ± 0.0	*	–	–	–	*
Weigela decora	0.2 ± 0.4 (1)	0.0 ± 0.0	0.0 ± 0.0	*	–	–	–	*
Fraxinus platypoda	–	–	–	*	0.8 ± 0.7 (4)	0.8 ± 0.8 (4)	0.4 ± 0.4 (2)	*
Ericaceae spp.	–	–	–		0.2 ± 0.3 (1)	0.0 ± 0.0	0.0 ± 0.0	
Unidentified species	0.0 ± 0.0	0.2 ± 0.4 (1)	0.0 ± 0.0		–	–	–	
Herbaceous plant								
Carex spp.	0.0 ± 0.0	12.4 ± 17.0 (62)	0.4 ± 0.5 (2)		0.4 ± 0.4 (2)	2.4 ± 2.8 (12)	0.0 ± 0.0	
Deinanthe bifida	0.0 ± 0.0[b]	0.0 ± 0.0[b]	3.8 ± 2.3[a] (19)	*	0.0 ± 0.0	0.0 ± 0.0	0.6 ± 0.5 (3)	*
Poaceae spp.	1.4 ± 1.7 (7)	1.0 ± 0.7 (5)	0.2 ± 0.4 (1)		0.0 ± 0.0	0.2 ± 0.3 (1)	0.4 ± 0.4 (2)	

Species												
Macleaya cordata	1.4 ± 1.1 (7)		0.4 ± 0.5 (2)		0.6 ± 0.5 (3)		–		–		–	
Compositae spp.	0.4 ± 0.5 (2)		0.6 ± 0.9 (3)		0.4 ± 0.9 (2)		–		–		–	
Mitella pauciflora	0.0 ± 0.0	*	0.0 ± 0.0	*	0.6 ± 0.9 (3)	*	–	*	–	*	–	*
Urtica laetevirens	0.2 ± 0.4 (1)	*	0.2 ± 0.4 (1)	*	0.2 ± 0.4 (1)	*	–	*	–	*	–	*
Laportea macrostachya	0.2 ± 0.4 (1)	*	0.2 ± 0.4 (1)	*	0.0 ± 0.0	*	0.0 ± 0.0	*	0.2 ± 0.3 (1)	*	0.0 ± 0.0	*
Chrysosplenium macrostemon	0.0 ± 0.0	*	0.0 ± 0.0	*	0.4 ± 0.9 (2)	*	–	*	–	*	–	*
Elatostema umbellatum var. majus	0.0 ± 0.0	*	0.0 ± 0.0	*	0.4 ± 0.5 (2)	*	–	*	–	*	–	*
Laportea bulbifera	0.0 ± 0.0	*	0.0 ± 0.0	*	0.4 ± 0.9 (2)	*	–	*	–	*	–	*
Stellaria diversiflora	0.2 ± 0.4 (1)	*	0.0 ± 0.0	*	0.0 ± 0.0	*	–	*	–	*	–	*
Viola spp.	–		–		–		0.6 ± 0.7 (3)		0.2 ± 0.3 (1)		0.0 ± 0.0	
Cardamine flexuosa	–		–		–		0.0 ± 0.0		0.4 ± 0.4 (2)		0.0 ± 0.0	
Chrysosplenium spp.	–		–		–		0.2 ± 0.3 (1)		0.0 ± 0.0		0.0 ± 0.0	
Scopolia japonica	–	*	–	*	–	*	0.0 ± 0.0	*	0.2 ± 0.3 (1)	*	0.0 ± 0.0	*
Scrophularia duplicato-serrata	–		–		–		0.0 ± 0.0		0.2 ± 0.3 (1)		0.0 ± 0.0	
Meehania urticifolia	–	*	–	*	–	*	0.0 ± 0.0	*	0.0 ± 0.0	=	0.2 ± 0.3 (1)	=
Unidentified species	–		–		–		0.2 ± 0.3 (1)		0.0 ± 0.0		0.0 ± 0.0	
Total number of seedlings	44.0 ± 43.9 (220)	*	32.6 ± 20.3 (163)	*	18.4 ± 5.1 (92)	*	18.0 ± 9.8 (90)	*	17.4 ± 6.4 (87)	*	6.4 ± 2.9 (32)	*
Total number of species	6.4 ± 2.1 (14)	*	7.4 ± 1.3 (15)	*	7.8 ± 1.1 (16)	*	7.4 ± 1.9 (12)	*	8.2 ± 2.5 (15)	*	4.8 ± 2.0 (9)	*

Species composition was determined from five soil samples, each 20 × 20 × 5 cm in depth, and five litter layer samples taken above the topsoil, and counting the number of seedlings emerging in each layer. Values indicate the mean number of seedlings emerging in each layer. Numbers in parentheses show the total number of seedlings recorded for each species. Letters indicate significant differences between the layers (Kruskal–Wallis, Sheffe test, $P < 0.05$). * indicates that the species occurred in the above-ground vegetation in each microtopography layer

Fig. 4.16 Seasonal change in *Cercidiphyllum japonicum* seedling survival rate under different relative light intensities at Ooyamazawa (Kubo et al. 2000, revised)

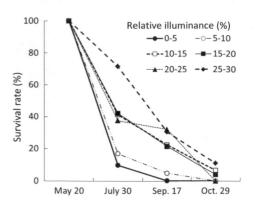

survival advantage in frequently disturbed riparian forests with steep slopes. Therefore, successful saplings are mostly likely to occur under bright light forest conditions (Kubo et al. 2000).

4.6 Sprouts

4.6.1 Structure of Multi-Stemmed Trunks

C. japonicum generally produces numerous sprouts (suckers), and it had the greatest mean number of stems (9.1 $\pm$ 12.2 per individual, $n = 59$) of all tree species in the Ooyamazawa study area (Table 4.1). Small sprout stems (shoot stems) occurred circularly around large (main) stems, although the distribution pattern of stems varied (Figs. 4.18 and 4.19). When considering stems arising from shoots, most *C. japonicum* stems are small in diameter ($n = 55$; Fig. 4.20), and individuals with larger main stems generally have a higher number of sprouts ($R = 0.37, n = 59$; Fig. 4.21; Kubo et al. 2001b). The mean number of sprouts per individual female, male, and immature tree was 8.9 $\pm$ 9.5, 12.5 $\pm$ 15.5, and 2.9 $\pm$ 4.4, respectively. Reproductive trees, both male and female, had a greater number of sprouts than did immature trees, although this difference was only significant between male and immature trees (Holm's method, $P < 0.05$). The sprout stem number was not significantly different between female and male trees (Holm's method, $P = 0.31$).

To clarify the dynamics of sprouts, we investigated age structure based on a growth-ring analysis of a *C. japonicum* stool (Kubo et al. 2005), which was harvested from between the slope and stream at the bottom of the study area valley. We cut all 29 associated stems at 50 cm high in autumn 2001. Across the study area, individual trees had large root systems from which many sprouts could originate. Sprouts frequently surrounded the main stems at the center of the stool (Fig. 4.22), and many individuals had coalesced stems (11 of 29 sprouts). The ages of coalesced stems tended to be similar (Fig. 4.22). Sprouts approximately 30 years old were usually clustered around main stems at the upstream slope site, whereas sprouts in

Fig. 4.17 Seedling height as influenced by light density (Kubo et al. 2004, revised). The upper and middle figures show the mean seedling height in their first year, November 1998, and their second year, January 2000. The lower figure shows the total dry weight of seedlings in the second year. Vertical bars indicate the standard deviation. Measurements were based only on seedlings that emerged in 1998, new seedlings germinating in 1999 were not included. Letters indicate significant differences between the light densities in each figure (*t*-test, $P < 0.05$). The level of significance was adjusted using the Holm's method. No seedlings survived the 3.0% RPPFD treatment over both years, and only one seedling remained in the 100.0% RPPFD treatment in the second year

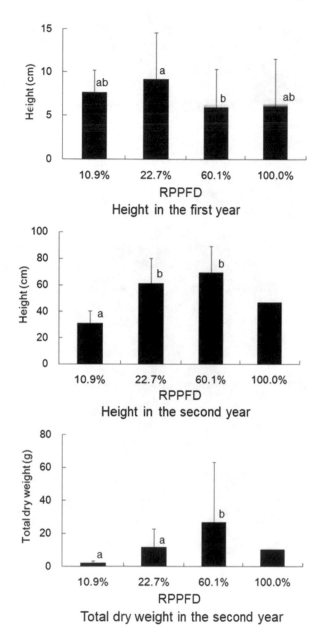

the downstream area tended to be older, around 80 years. Eighty-year-old sprouts from the slope site showed increased growth approximately 30 years ago, as determined by growth-ring analysis (Fig. 4.23). This suggests that light conditions may have improved around that time, allowing many shoots to sprout or grow rapidly, and potentially increasing stem coalescence. Growth patterns varied by

Fig. 4.18 Various multi-stemmed trunks (stools) of *Cercidiphyllum japonicum* in Ooyamazawa

individual and with age (Fig. 4.23) and sprout stem diameter was positively correlated with age ($R^2 = 0.66$, $n = 45$; Fig. 4.24).

Most stems cut in 2001 produced new sprouts from their stools the following
September (Fig. 4.25). The number of current-year sprouts was positively correlated
with the age and diameter of the parent stems (Kubo et al. 2005). Smaller, younger
stems also produced new sprouts, and new sprouts would survive on the periphery of
the stand under favorable light conditions.

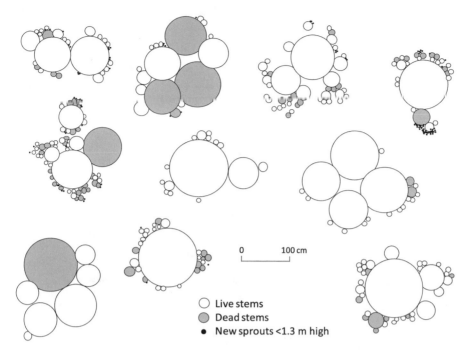

Fig. 4.19 Distribution patterns of stems for 10 typical, large *Cercidiphyllum japonicum* individuals in Ooyamazawa

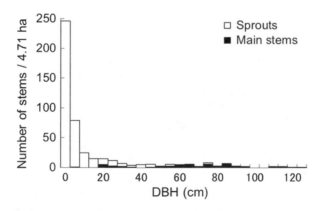

Fig. 4.20 DBH size class distribution, including main stems and sprouts, of *Cercidiphyllum japonicum* (Kubo et al. 2001b, revised)

4.6.2   Self-Maintenance by Sprouting

We show a proposed scheme for self-maintenance of *C. japonicum* by sprouting in Fig. 4.26. Following germination and growth, sprouts are produced as a result of

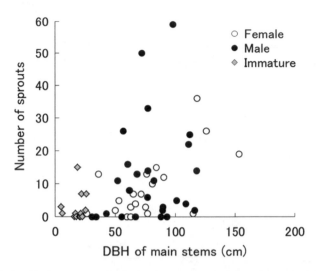

Fig. 4.21 Relationship between the DBH of main stems and number of sprouts of *Cercidiphyllum japonicum* (Kubo et al. 2001b, revised)

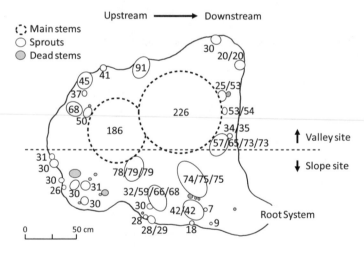

Fig. 4.22 Distribution of main stems and sprouts, denoted by age, of *Cercidiphyllum japonicum* (Kubo et al. 2005, revised). The upper site was in the valley, and the lower site was on the slope, as indicated

endogenous factors, such as aging, or in response to external factors, such as gap formation and physical damage. Following stem death, *C. japonicum* is able to fill in ensuing gaps by sprouting. Consequently, colonies containing sprouts of various ages are produced, which are circularly distributed around the stool. By this process, broad, extensive canopies and multi-stemmed individuals can be produced from a single main stem.

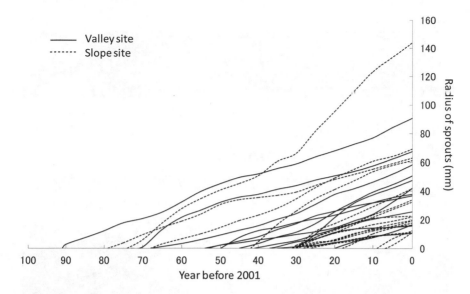

Fig. 4.23 Radial growth in *Cercidiphyllum japonicum* sprouts (Kubo et al. 2005, revised). Valley and slope sites are as indicated in Fig. 4.22

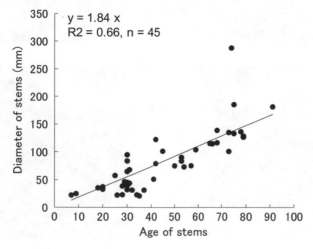

Fig. 4.24 Relationship between stem age and diameter (Kubo et al. 2005, revised). Age and diameter were measured 50 cm from the ground

4.6.3   Sprouting Traits of C. japonicum and C. magnificum

On the upper stream talus slope of Ooyamazawa (Fig. 4.6), almost all *C. japonicum* and *C. magnificum* produced numerous sprouts (Fig. 4.27); *C. magnificum* had a greater number of smaller stems and lower number of large stems relative to

Fig. 4.25 Current-year sprouts from logged stems. (**a**) Bridge made from logged stems of *Cercidiphyllum japonicum*; (**b**) and (**c**) new sprouts arising from logged stumps

C. japonicum (Fig. 4.28). The average DBH of the main stems of *C. japonicum* was significantly larger than that of *C. magnificum* (*t*-test, $P < 0.01$). A number of *C. magnificum* individuals only reached the subcanopy layer, despite their being mature (Kubo et al. 2010). Dead stems in this species were numerous and often

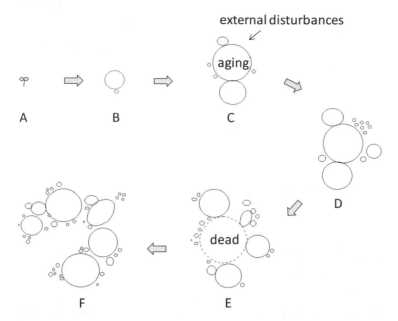

Fig. 4.26 Proposed scheme for the self-maintenance of *Cercidiphyllum japonicum* by sprouting (Kubo et al. 2005, revised). *Cercidiphyllum japonicum* germinates (**a**), grows to approximately 40 years of age (**b**), then produces sprouts as a function of aging at approximately 130 years (**c**), and in response to long-term external disturbances at approximately 230 years (**d**). Following the death of the parent stem, ensuing gaps fill with new sprouts at approximately 300 years (**e**). Consequently, *C. japonicum* creates colonies composed of sprouts of various ages (**f**). Age estimates were obtained from the individual shown in Fig. 4.22

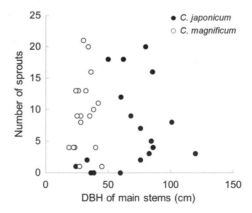

Fig. 4.27 Relationship between the DBH of main stems and number of sprouts of *Cercidiphyllum japonicum* and *Cercidiphyllum magnificum*, co-occurring on an upper stream talus slope at Ooyamazawa

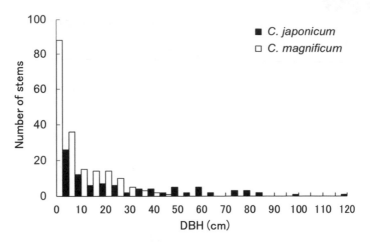

Fig. 4.28 DBH size class distribution of stems including main stems and sprouts of *Cercidiphyllum japonicum* and *Cercidiphyllum magnificum*, co-occurring on an upper stream talus slope at Ooyamazawa (Kubo et al. 2010, revised)

>30 cm DBH; the maximum recorded diameter of a live stem was 45 cm (Fig. 4.27). This suggests that stems are pressured to reach optimal size to ensure survival.

  Cercidiphyllum are distributed in the montane and subalpine zones in Japan, and differences in stool structure between *C. japonicum* and *C. magnificum* reflect differences in the climatic conditions these species experience. Severe conditions such as wind and snowfall prevail in the subalpine zone in Japan (Shidei 1956; Yoshino 1973). One response to high disturbance severity and frequency is to reduce above-ground biomass and adopt a multi-stemmed, re-sprouting architecture (Bellingham and Sparrow 2000). Subalpine zone conditions may therefore require self-maintenance by *C. magnificum* by producing numerous sprouts with high mortality rates. In contrast, larger, taller individuals of *C. japonicum* indicate exposure to competition with co-occurring species for light in the canopy layer.

4.7   Conclusions

Cercidiphyllum japonicum reaches reproductive maturity at approximately 30 cm DBH (Fig. 4.9), which is estimated to occur after 100 years of growth. Larger individuals with many large sprouts flower heavily in early spring and then disperse a large amount of winged seeds following annual seasonal defoliation in autumn (Figs. 4.12 and 4.13). Suitable germination sites for *C. japonicum* seedlings include bare soil and fallen trees; sites with high litter accumulation or gravel are unsuitable for germination (Figs. 4.14 and 4.15) due to the small size of seeds and seedlings (Figs. 4.10 and 4.14). Germination does not imply survival for this species, given the high observed first-year mortality, which was likely a result of desiccation or stream flow and precipitation events (Fig. 4.16). Larger seedlings growing under bright light conditions may have a survival advantage (Fig. 4.17).

Despite the expansive canopies and rootstocks of parent trees, saplings of *C. japonicum* are uncommon (Fig. 4.4). This species occurs at low density in its riparian habitat (Fig. 4.5), but produces many small stems per individual (Figs. 4.20 and 4.21). Sprouts are continually produced as a result of endogenous and external factors, and *C. japonicum* can self-maintain for several hundreds of years by sprouting (Fig. 4.26), potentially compensating for low sapling recruitment with high vegetative reproduction.

The oldest specimen of *F. platypoda*, a coexisting canopy species, is 254 years (Sakio 1997), and the lifespan of *P. rhoifolia* is approximately 120 years (Kisanuki et al. 1992). We found that *C. japonica* stands had live main stems dating to 226 years (Fig. 4.22). Many stools in Ooyamazawa contain the remains of previous main stems, meaning that many individuals are likely several hundred years old. It is possible, therefore, that *C. japonica* maintains its populations by outliving its competitors.

References

Bellingham PJ, Sparrow AD (2000) Resprouting as a life history strategy in woody plant communities. Oikos 89:409–416

Crane PR, Stockey RA (1985) Growth and reproductive biology of *Joffrea speirsii* gen. et sp. nov., a *Cercidiphyllum*-like plant from the Late Paleocene of Alberta, Canada. Can J Bot 63:340–364

Dosmann MS (1999) Katsura: a review of *Cercidiphyllum* in cultivation and in the wild. New Plantsman 6:52–62

Farmer RE (1997) Seed ecophysiology of temperate and boreal zone forest trees. St. Lucie Press, Delray Beach

Isagi Y, Kudo M, Osumi K, Sato T, Sakio H (2005) Polymorphic microsatellite DNA markers for a relictual angiosperm *Cercidiphyllum japonicum* and their utility for *Cercidiphyllum magnificum*. Mol Ecol Notes 5:596–598

Kawanishi M, Sakio H, Ohno K (2004) Forest floor vegetation of *Fraxinus platypoda - Pterocarya rhoifolia* forest along Ooyamazawa valley in Chichibu, Kanto district, Japan, with a special reference to ground disturbance. Veg Sci 21:15–26. (in Japanese with English abstract)

Kayane I (1980) Physical geography course volume 5, hydrology. Daimei-do, Tokyo (in Japanese)

Kisanuki H, Kaji M, Suzuki K (1992) Structure and regeneration process of ash (*Fraxinus spaethiana* Ling.) stands in Chichibu Mountains. Bull Tokyo Univ For 88:15–32. (in Japanese with English summary)

Kubo M, Shimano K, Sakio H, Ohno K (2000) Germination sites and establishment conditions of *Cercidiphyllum japonicum* seedlings in the riparian forest. J Jpn For Soc 82:349–354. (in Japanese with English summary)

Kubo M, Shimano K, Ohno K, Sakio H (2001a) Relationship between habitats of dominant trees and vegetation units in Chichibu Ohyamasawa riparian forest. Veg Sci 18:75–85. (in Japanese with English summary)

Kubo M, Shimano K, Sakio H, Ohno K (2001b) Sprout trait of *Cercidiphyllum japonicum* based on the relationship between topographies and sprout structure. J Jpn For Soc 83:271–278. (in Japanese with English summary)

Kubo M, Sakio H, Shimano K, Ohno K (2004) Factors influencing seedling emergence and survival in *Cercidiphyllum japonicum*. Folia Geobot 39:225–234

Kubo M, Sakio H, Shimano K, Ohno K (2005) Age structure and dynamics of *Cercidiphyllum japonicum* sprouts based on growth ring analysis. For Ecol Manag 213:253–260

Kubo M, Kawanishi M, Shimano K, Sakio H, Ohno K (2008) The species composition of soil seed banks in the Ooyamazawa riparian forest, in the Chichibu Mountains, Central Japan. Jpn For Soc 90:121–124. (in Japanese with English summary)

Kubo M, Shimano K, Sakio H, Isagi Y, Ohno K (2010) Difference between sprouting traits of *Cercidiphyllum japonicum* and *C. magnificum*. J For Res 15:337–340

Li J, Dosmann MS, Tredici PD (2002) Systematic relationship of weeping Katsura based on nuclear ribosomal DNA sequences. Hort Sci 37(3):595–598

Manchester SR, Chen ZD, Lu AM, Uemura K (2009) Eastern Asian endemic seed plant genera and their paleogeographic history throughout the Northern Hemisphere. J Syst Evol 47:1–42

Mizui N (1993) Ecological studies on reproduction in deciduous broad-leaved tree species. Bull Hokkaido For Res Inst 30:1–67. (in Japanese)

Ohno K (2008) Vegetation-geographic evaluation of the syntaxonomic system of valley-bottom forests occurring in the cool-temperate zone of the Japanese Archipelago. In: Sakio H, Tamura T (eds) Ecology of riparian forests in Japan. Springer, Tokyo, pp 49–72

Qi X-S, Chen C, Comes HP, Sakaguchi S, Liu Y-H, Tanaka N, Sakio H, Qiu Y-X (2012) Molecular data and ecological niche modelling reveal a highly dynamic evolutionary history of the East Asian Tertiary relict *Cercidiphyllum* (Cercidiphyllaceae). New Phytol 196:617–630

Sakio H (1997) Effects of natural disturbance on the regeneration of riparian forests in a Chichibu mountains, central Japan. Plant Ecol 132:181–195

Sakio H, Kubo M, Shimano K, Ohno K (2002) Coexistence of three canopy tree species in a riparian forest in the Chichibu mountains, Central Japan. Folia Geobot 37:45–61

Seiwa K, Kikuzawa K (1996) Importance of seed size for the establishment of seedlings of five deciduous broad-leaved tree species. Vegetatio 123:51–64

Shidei T (1956) A concept about the reason why coniferous forest zone is partially lacking in sub-alpine zone along the sea of Japan. J Jpn For Soc 38:356–358. (in Japanese)

Skawińska K (1986) Some new and rare pollen grains from Neogene deposits at Ostrzeszów (South-West Poland). Acta Palaeobotanica 25:107–118

Thomas P (2000) Trees: their natural history. Cambridge University Press, Cambridge (Kumazaki M, Asakawa S, Sudou S (2001) Trees: their natural history. Tsukiji-shokan, Tokyo (in Japanese))

Yoshino M (1973) Studies on wind-shaped trees: their classification, distribution and significance as a climatic indicator. Climatological Notes 12:1–52

Chapter 5
Acer Tree Species

Masako Kubo, Hitoshi Sakio, Motohiro Kawanishi, and Motoki Higa

Abstract Thirteen *Acer* species were found in Ooyamazawa riparian forest in Japan. Of these, the shrubby species *A. carpinifolium* and *A. argutum* and the tall tree species *A. shirasawanum* and *A. mono* were the most abundant. Sprouts of shrubby species were common, but not of tall tree species, which implies that individuals of shrubby species are maintained in the understory layer via sprouts, whereas tall tree species extend upward toward the canopy. All four *Acer* species were found mainly on upstream sediments; moreover, the density of *A. shirasawanum* was high, implying that *A. shirasawanum* may eventually become an important species in the sedimentary upstream area. On the other hand, *A. carpinifolium* was dominant in the unstable downstream V-shaped valley; this species adapts to the disturbed downstream area by producing more sprouts. These differences in life history promote diversity in the forest structure of Ooyamazawa riparian forest.

Keywords *Acer argutum* · *Acer carpinifolium* · *Acer pictim* · *Acer shirasawanum* · Distribution · Riparian forest · Sprouting traits

M. Kubo (✉)
Faculty of Life and Environmental Science, Shimane University, Shimane, Japan
e-mail: kubom@life.shimane-u.ac.jp

H. Sakio
Sado Island Center for Ecological Sustainability, Niigata University, Niigata, Japan
e-mail: sakio@agr.niigata-u.ac.jp; sakiohit@gmail.com

M. Kawanishi
Faculty of Education, Kagoshima University, Kagoshima, Japan
e-mail: kawanishi@edu.kagoshima-u.ac.jp

M. Higa
Faculty of Science and Technology, Kochi University, Kochi, Japan
e-mail: mhiga@kochi-u.ac.jp

© The Author(s) 2020
H. Sakio (ed.), *Long-Term Ecosystem Changes in Riparian Forests*, Ecological Research Monographs, https://doi.org/10.1007/978-981-15-3009-8_5

5.1   Introduction

The genus *Acer* is comprised of tree species that play important roles in forests across the Northern Hemisphere, particularly in late-successional forests dominated by hardwoods such as beech (*Fagus*) species (Runkle 1990; Poulson and Platt 1996; Cao and Ohkubo 1999). *Acer* species often coexist with other deciduous species, depending on light conditions and/or disturbance regimes (Sipe and Bazzaz 1995). Some *Acer* species are found in riparian forests (Ohno 2008), in topographical habitats created by various types of disturbance (Masaki et al. 1992, 2005; Suzuki et al. 2002), suggesting that some *Acer* species may exhibit pioneer traits.

In Japan, 28 *Acer* tree species are distributed throughout sub-tropical, warm-temperate, cool-temperate, and sub-alpine forests (Yonekura 2012). Many trees in riparian forests of the cool-temperate zone in Japan are *Acer* species (Ohno 2008). *Acer* is therefore an important taxonomic group that contributes to tree species diversity and forest structure in these riparian forests. *Acer* species typically coexist with other tree species due to their different life histories (Sakio et al. 2002), and represent both early successional species, in response to various riparian disturbances, and late-successional species, which play important roles in riparian forest dynamics.

Acer carpinifolium Siebold et Zucc., *Acer shirasawanum* Koidz., *Acer pictum* Thunb., *Acer argutum* Maxim., *Acer nipponicum* H.Hara, *Acer rufinerve* Siebold et Zucc., *Acer tenuifolium* (Koidz.) Koidz., *Acer palmatum* Thunb., *Acer amoenum* Carrière var. *amoenum*, *Acer maximowiczianum* Miq., *Acer cissifolium* (Siebold et Zucc.) K.Koch, *Acer distylum* Siebold et Zucc., and *Acer micranthum* Siebold et Zucc. all grow in the Ooyamazawa riparian area of central Japan, and the four dominant *Acer* species are *Acer carpinifolium*, *A. shirasawanum*, *A. pictim*, and *A. argutum*. In this chapter, we clarify the life histories of these four major *Acer* species and discuss the relationship between their life histories and riparian topography in Ooyamazawa.

5.2   Study Species

A. carpinifolium, *A. shirasawanum*, *A. pictim*, and *A. argutum* are indigenous to Japan. *A. carpinifolium* and *A. pictim* are distributed in Honshu, Shikoku, and Kyusyu Islands, and *A. shirasawanum* and *A. argutum* are distributed in Honshu and Shikoku Islands. *A. carpinifolium* and *A. argutum* generally grow to shrub height (Table 5.1, Fig. 5.1). These four *Acer* species are deciduous and monophyllous; *A. carpinifolium* has oval leaves, whereas the remaining three *Acer* species have palmate leaves.

A. carpinifolium is a dioecious shrub that grows to a height of approximately 10 m and produces shoots around stems. This species is mainly distributed in riparian areas in the mountains, flowering in May and producing seeds during

Table 5.1 Traits of four *Acer* tree species found in Ooyamazawa

	A. carpinifolium	A. shirasawanum	A. pictim	A. arguturа
Tree height	Shrubby	Tall	Tall	Shrubby
Sexual expression	Dioecy	Monoecy	Monoecy	Dioecy
Flower season	May	May	May	May
Fruit season	Sep.–Oct.	Sep.–Oct.	Sep.–Oct.	Sep.–Oct.
Sprout	Around stem	Few	Few	Around stem and root sucker

Fig. 5.1 Tree form of the four most common *Acer* species in the Ooyamazawa riparian site. (**a**) *A. shirasawanum* in the subcanopy layer. (**b**) *A. carpinifolium* in the shrub layer. (**c**) *A. argutum* in the shrub layer. (**d**) *A. pictim* in the subcanopy layer

September–October. *A. shirasawanum* is a monoecious tree that grows to a height of about 20 m and diameter at breast height (DBH) of 30–40 cm, with larger individuals reaching a DBH of 80 cm. *A. shirasawanum* is mainly found in mountain areas; it flowers in May and produces seeds during September–October. *A. pictim* is a monoecious tree that grows to a height of 20 m and DBH of 50–60 cm; it is found in riparian and mountain areas, flowers in May, and produces seeds during September–October. *A. argutum* is a dioecious shrub that grows to a height of about 8 m and produces shoots close to the root system; it is found in riparian and mountain areas, flowers in May, and produces seeds during September–October.

5.3 Structure and Distribution

5.3.1 *Abundance and Structure of Four* Acer *Species*

In this study, we identified 1204 individual trees (255.4 trees/ha) of 13 *Acer* species in a 4.71-ha plot within the Ooyamazawa riparian forest (Table 5.2, Fig. 5.2). Thus, Ooyamazawa contains roughly half of the 28 *Acer* species found in Japan (Yonekura 2012). *Acer* species comprised 28.9% of a total of 45 tree species and 57.0% of a total of 2111 individuals (448.1/ha) in the study area. Together, *A. carpinifolium*, *A. shirasawanum*, *A. pictum*, and *A. argutum* comprised 96.3% of all *Acer* individuals and 55.0% of all individual trees.

Acer species densities were particularly high in the subcanopy and shrub layers (Table 5.2, Fig. 5.3). *A. carpinifolium* and *A. argutum* were found in both the subcanopy and shrub layers; however, the vast majority of individuals were found in the shrub layer. In contrast, *A. shirasawanum* and *A. pictum* were found in all layers, with most individuals in the subcanopy and shrub layers.

Small *Acer* individuals (DBH < 40 cm) were numerous; most individual *A. carpinifolium* and *A. argutum* shrubs did not exceed a DBH of 20 cm (Fig. 5.4). *A. pictum* had the largest DBH at 92.0 cm, followed by *A. shirasawanum* at 62.8 cm (Table 5.2).

5.3.2 *Spatial Distribution of Four* Acer *Species*

The four *Acer* species were mainly distributed on upstream sedimentary debris flows; among these, only *A. carpinifolium* was also dominant in the downstream V-shaped valley (Fig. 5.5). We divided the study plot into 20 subplots along the stream and compared the densities of the four *Acer* species and the dominant canopy tree *Fraxinus platypoda* in the canopy, subcanopy, and shrub layers in each subplot (Fig. 5.6). In the subcanopy layer, *A. carpinifolium* density was higher in the downstream valley (Wilcoxon rank sum test, $P < 0.01$), although in the shrub layer *A. carpinifolium* density was high both upstream and downstream (Wilcoxon rank sum test, $P = 0.16$). In the subcanopy layer, *A. argutum* density was low, while in the shrub layer, *A. argutum* density was significantly higher upstream than downstream in the valley (Wilcoxon rank sum test, $P < 0.01$). In both the subcanopy and shrub layers, *A. shirasawanum* density was higher upstream (Wilcoxon rank-sum test, $P < 0.01$), as was *A. pictum* density in the shrub layer (Wilcoxon rank sum test, $P < 0.01$), although *A. pictum* densities in the canopy and subcanopy layers did not differ significantly between upstream and downstream areas (Wilcoxon rank sum test, canopy layer; $P = 0.28$, subcanopy layer; $P = 0.12$).

The sedimentary upstream area was the best habitat for three of the four *Acer* species, whereas *A. carpinifolium* was better suited to the downstream valley (Fig. 5.5). The upstream alluvial fan and terrace debris flows contain rich soil and

Table 5.2 Tree density of each *Acer* species found in Ooyamazawa

| | Number (/ ha) | Number(/ha) | | | Maximum DBH (cm) | Mean DBH (cm) | Mean number of shoot stems |
		Canopy	Subcanopy	Shrub			
A. carpinifolium	92.1	–	5.1	87.0	22.0	8.0 ± 3.2	6.0 ± 4.3
A. shirasawanum	78.5	0.4	27.0	51.2	62.8	13.2 ± 8.7	0.2 ± 0.6
A. pictim	57.1	2.8	20.8	33.5	92.0	13.3 ± 10.7	0.1 ± 0.4
A. argutum	18.5	–	0.4	18.0	13.2	6.4 ± 2.2	5.3 ± 4.1
A. nipponicum	2.5	0.2	0.6	1.7	28.7	10.0 ± 7.3	2.1 ± 2.5
A. rufinerve	2.3	0.2	1.1	1.1	33.6	12.1 ± 7.6	0.2 ± 0.4
A. tenuifolium	1.5	–	0.4	1.1	15.0	8.1 ± 3.0	0.4 ± 0.7
A. palmatum	1.3	–	0.6	0.6	23.9	14.1 ± 6.1	0.0
A. amoenum var. *moenum*	0.6	–	0.6	–	15.5	11.2 ± 3.2	0.0
A. maximowiczianum	0.4	–	0.4	–	39.0	28.1 ± 11.0	0.5 ± 0.5
A. cissifolium	0.2	–	0.2	–	18.2	18.2	3.0
A. distylum	0.2	–	0.2	–	14.0	14.0	2.0
A. micranthum	0.2	–	–	0.2	4.1	4.1	1.0
Acer species	255.4	3.6	57.5	194.5	92.0	10.8 ± 7.9	2.7 ± 4.0
Total species	448.1	104.0	99.6	244.6	153.4	21.4 ± 22.4	2.0 ± 4.1

Values are means ± standard deviation (SD) for each species. *DBH* Diameter at breast height

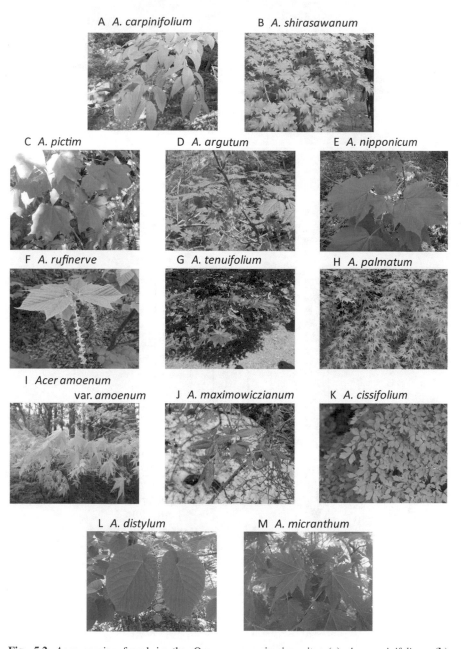

Fig. 5.2 *Acer* species found in the Ooyamazawa riparian site. (**a**) *A. carpinifolium*. (**b**) *A. shirasawanum*. (**c**) *A. pictim*. (**d**) *A. argutum*. (**e**) *A. nipponicum*. (**f**) *A. rufinerve*. (**g**) *A. tenuifolium*. (**h**) *A. palmatum*. (**i**) *A. amoenum* var. *amoenum*. (**j**) *A. maximowiczianum*. (**k**) *A. cissifolium* (photo by Takuto Shitara). (**l**) *A. distylum*. (**m**) *A. micranthum*

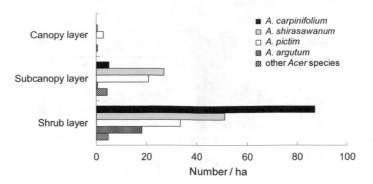

Fig. 5.3 Tree density among *Acer* species for each layer. Trees in the canopy layer reach the canopy and exceed a height of 20 m and DBH of 20 cm. Trees in the subcanopy layer do not reach the canopy and are <20 m high. Trees in the shrub layer are <10 m high

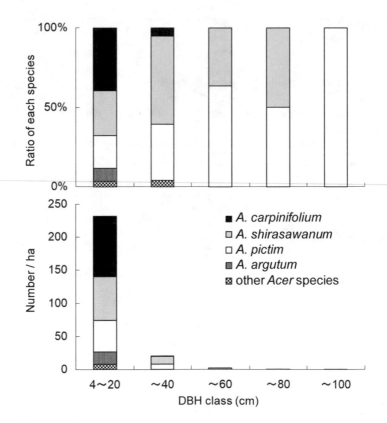

Fig. 5.4 DBH class distribution of *Acer* species. Upper, DBH ratio for each *Acer* species; lower, DBH frequency distribution for *Acer* species

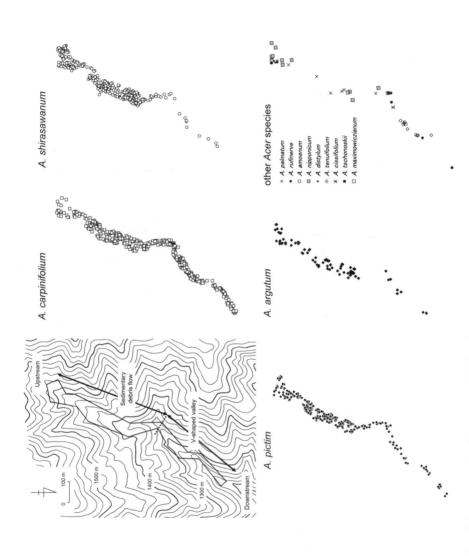

Fig. 5.5 Distribution of *Acer* trees at the Ooyamazawa riparian site

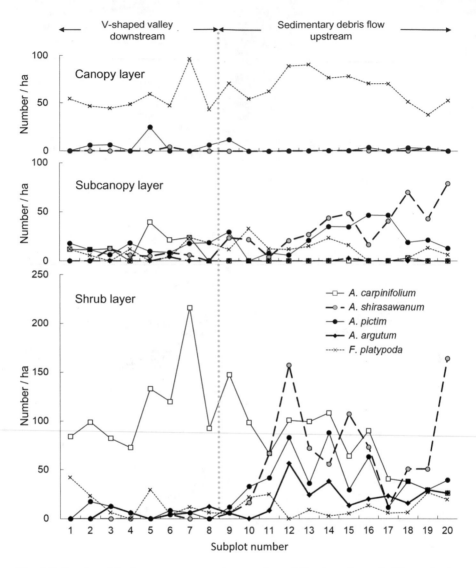

Fig. 5.6 Density of the four most common *Acer* species and *Fraxinus platypoda*. The study plot was divided into 20 subplots at 60-m intervals along the stream, with eight subplots (1–8) in the downstream V-shaped valley and 12 subplots (9–20) in the upstream sedimentary debris flow. Upper, tree density in the canopy layer; middle, tree density in the subcanopy layer; lower, tree density in the shrub layer

a considerable litter layer; in contrast, little litter is found on terrace scarps, new landslide sites, old landslide slopes, or talus of the downstream V-shaped valley (Kawanishi et al. 2004). Downstream disturbances including erosion and sedimentation of soil, sand, and/or gravel are frequent due to stream flow and/or steep slopes.

Fewer disturbances occur upstream where slopes are gentle (about 12°) and the valley bottom has been filled by large landslides and/or debris flows.

Factors determining the distribution patterns of *Acer* species can include shade tolerance, the distance from the seed source, and germination site conditions, e.g., areas of soil and litter accumulation. Leaf litter cover also reduces the risk of predation on *Acer* seeds (Tanaka 1995). Although *A. pictum* was found in the canopy and subcanopy layers both upstream and downstream, *A. shirasawanum* clearly occurs in the subcanopy and shrub layers only upstream, where it is dominant (Figs. 5.5 and 5.6). *A. pictum* saplings can survive even in the forest understory (Abe et al. 1995; Hara 1987; Masaki et al. 1992) by acclimating to deep shade (Kitao et al. 2006). *A. shirasawanum* can regenerate in smaller gaps than *A. pictum*, due to its shade tolerance (Sakai 1986). It remains unclear why the range of *A. shirasawanum* does not extend downstream, since its shade tolerance should allow it to dominate the more stable upstream sediments, where soils are rich, litter accumulation is greater, and the upland forest is a nearby seed source. For these reasons, *A. shirasawanum* may eventually become an important species in the sedimentary upstream area.

5.4 Sprouting Traits

The two *Acer* shrub species, *A. carpinifolium* and *A. argutum*, produced large numbers of shoots (Fig. 5.7). Among *Acer* species, the greatest mean number of shoots was observed in *A. carpinifolium* (6.0 ± 4.3), exceeded only by *Cercidiphyllum japonicum* (9.0 ± 12.1) in the Ooyamazawa study site (Chap. 4). *A. shirasawanum* and *A. pictum*, both of which are tall tree species, had fewer shoots, suggesting that *A. carpinifolium* and *A. argutum* may reproduce in the shrub layer via sprouting.

Shrub species tend to produce many shoots (Midgley 1996); reproduction via sprouting provides an advantage in habitats where environmental conditions are severe (Sakai et al. 1995; Kubo et al. 2010). The low light conditions in the shrub layer prevent *A. carpinifolium* and *A. argutum* from consistently receiving direct light. Main shoots tend to die when they grow too large to balance photosynthesis and respiration, allowing a large shoot with large leaf area to become a new main shoot.

The large numbers of shoots observed in the shrub *A. carpinifolium* are therefore suitable for habitats with significant surface erosion, allowing long-term survival of some individuals in the subcanopy layer within the downstream V-shaped valley, where the density of other *Acer* species is low (Figs. 5.5 and 5.6). Similarly, *Euptelea polyandra* has shoots adapted to its unstable habitat conditions (Sakai et al. 1995), with main shoots gradually inclining as they increase in size, facilitating the establishment of younger shoots. The large number of shoots and dominance of *A. carpinifolium* in the unstable V-shaped valley suggest that this species has sprouting traits adapted to the unstable steep slopes and disturbance regime of this site.

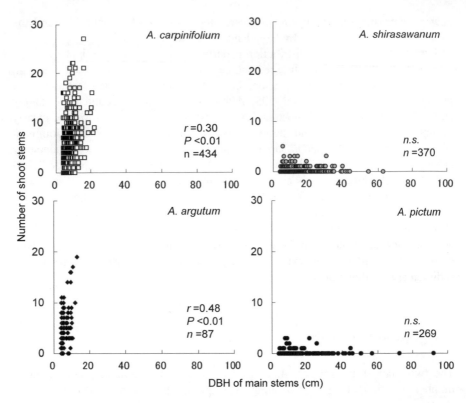

Fig. 5.7 Relationship between shoot number and DBH in the four most common *Acer* species

Some species reproduce via spontaneous sprouting (Verwijst 1988; Keeley 1992). In this study, almost all *A. argutum* were found on stable upstream sediments. Many *A. argutum* shoots are produced near the soil around the main stem, like root suckers, whereas *A. carpinifolium*, which was dominant in the unstable V-shaped valley, sprouts from the shoot base. Some tree species extend their roots to the surface to produce new shoots (Gyokusen et al. 1991; Ogawa et al. 1999; Sakio 2015). Therefore *A. argutum* reproduces through sprouting in the shrub layer on stable upstream sediments, where soil is rich, and its range does not extend to the V-shaped valley, where slopes are steep and soil is poor.

5.5   Conclusion

In this study, we identified 13 *Acer* tree species in a 4.71-ha study plot in the Ooyamazawa riparian forest. Among these, four species, *A. carpinifolium*, *A. shirasawanum*, *A. pictum*, and *A. argutum* constituted 55.0% of all individuals. These four coexisting *Acer* species had different life history characteristics in terms of size, distribution, and shoot production. *A. shirasawanum* and *A. pictum* are tall tree species, whereas *A. carpinifolium* and *A. argutum* are shrub species (Table 5.2, Fig. 5.3), which produce more shoots than the taller species (Fig. 5.7). All four *Acer* species were mainly found on upstream sediments. However, *A. shirasawanum* and *A. argutum* were distributed only on stable upstream sediments, *A. pictum* was distributed in the canopy and subcanopy layers in both areas, and *A. carpinifolium* was dominant in the unstable downstream V-shaped valley (Figs. 5.5 and 5.6). The two shrub species have different sprouting traits, with *A. carpinifolium* adapting to the disturbed downstream area by producing more shoots and *A. argutum* producing a lot of shoots upstream, where the soil is rich and disturbance occurs less frequently. These differences in life history promote diversity in forest structure in this riparian forest.

References

Abe S, Masaki T, Nakashizuka T (1995) Factors influencing sapling composition in canopy gaps of a temperate deciduous forest. Vegetatio 120:21–32

Cao K-F, Ohkubo T (1999) Suppression and release during canopy recruitment in *Fagus crenata* and *Acer mono* in two old-growth beech forests in Japan. Plant Ecol 145:281–290

Gyokusen K, Iijima Y, Yahata H (1991) Spacial distribution and morphological features of root sprouts in Niseakasia (Robinia pseudo-acasia L.) growing under a coastal black pine forest. Bull Kyushu Univ For 64:13–28 (Japanese with English summary)

Hara M (1987) Analysis of seedling banks of a climax beech forest: ecological importance of seedling sprouts. Vegetatio 71:67–74

Kawanishi M, Sakio H, Ohno K (2004) Forest floor vegetation of *Fraxinus platypoda—Pterocarya rhoifolia* forest along Ooyamazawa valley in Chichibu, Kanto district, Japan, with a special reference to ground disturbance. Veg Sci 21:15–26. (In Japanese with English abstract)

Keeley JE (1992) Recruitment of seedlings and vegetative sprouts in unburned chaparral. Ecology 73:1194–1208

Kitao M, Lei TT, Koike T, Tobita H, Maruyama Y (2006) Tradeoff between shade adaptation and mitigation of photoinhibition in leaves of *Quercus mongolica* and *Acer mono* acclimated to deep shade. Tree Physio 26:441–448

Kubo M, Shimano K, Sakio H, Isagi Y, Ohno K (2010) Difference between sprouting traits of *Cercidiphyllum japonicum* and *C. magnificum*. J For Res 15:337–340

Masaki T, Osumi K, Takahashi K, Hoshizaki K (2005) Seedling dynamics of Acer mono and Fagus crenata: an environmental filter limiting their adult distributions. Plant Ecol 177:189–199

Masaki T, Suzuki W, Niiyama K, Iida S, Tanaka H, Nakashizuka T (1992) Community structure of a species-rich temperate forest, Ogawa Forest Reserve, central Japan. Vegetatio 98:97–111

Midgley JJ (1996) Why the world's vegetation is not totally dominated by resprouting plants; because resprouters are shorter than reseeders. Ecography 19:92–95

Ogawa M, Aiba Y, Watanabe N (1999) Morphological and anatomical characteristics of sprouting root of *Prunus ssiori*. J Jpn For Soc 81:36–41

Ohno K (2008) Vegetation-geographic evaluation of the syntaxonomic system of valley-bottom forests occurring in the cool-temperate zone of the Japanese Archipelago. In: Sakio H, Tamura T (eds) Ecology of riparian forests in Japan. Springer, Tokyo, pp 49–72

Poulson TL, Platt WJ (1996) Replacement patterns of beech and sugar maple in warren woods, Michigan. Ecology 77:1234–1253

Runkle JR (1990) Gap dynamics in an Ohio *Acer-Fagus* forest and speculations on the geography of disturbance. Can J For Res 20:632–641

Sakai A, Ohsawa T, Ohsawa M (1995) Adaptive significance of sprouting of *Euptelea polyandra*, a deciduous tree growing on steep slopes with shallow soil. J Plant Res 108:377–386

Sakai S (1986) Patterns of branching and extension growth of vigorous saplings of Japanese *Acer* species in relation to their regeneration strategies. Can J Bot 65:1578–1585

Sakio H, Kubo M, Shimano K, Ohno K (2002) Coexistence of three canopy tree species in a riparian forest in the Chichibu mountains, central Japan. Folia Geobotanica 37:45–61

Sakio H (2015) Why did the black locust expand broadly at the river basin in Japan? Jpn Soc Reveg Technol 40(3):465–471 (Japanese with English abstract)

Sipe TW, Bazzaz FA (1995) Gap partitioning among maples (*Acer*) in central New England: survival and growth. Ecology 76:1587–1602

Suzuki W, Osumi K, Masaki T, Takahashi K, Daimaru H, Hoshizaki K (2002) Disturbance regimes and community structures of a riparian and an adjacent terrace stand in the Kanumazawa Riparian Research Forest, northern Japan. For Ecol Manage 157:285–301

Tanaka H (1995) Seed demography of three co-occurring *Acer* species in a Japanese temperate deciduous forest. J Veg Sci 6:887–896

Verwijst T (1988) Environmental correlates of multiple-stem formation in *Betula pubescens* ssp. *tortuosa*. Vegetatio 76:29–36

Yonekura K (2012) An enumeration of the vascular plants of Japan. Hokuryukan, Tokyo

Part III
Diversity and Coexistence in Riparian Forests

Part II

Chapter 6
Diversity of Herbaceous Plants in the Ooyamazawa Riparian Forest

Motohiro Kawanishi

Abstract Various herbaceous plants grow in the forest floor of the Ooyamazawa riparian forest. The diversity of herbs is related to the complexity of the ground surface condition, which is formed by ground disturbances such as debris flow, landslides, and soil erosion. Most notably, micro-scale heterogeneity and disturbances have effects on the growth of herbs. Herbaceous plants may adapt to such ground conditions throughout their life cycle, i.e., during vegetative growth, vegetative reproduction, and sexual reproduction. We can observe a part of these ecological characteristics as functional groups. Furthermore, we will show the relationships between the ecological functional traits and their relation to vegetative reproduction and micro-disturbances in riparian areas.

Keywords *Chrysosplenium macrostemon* · *Deinanthe bifida* · *Elatostema umbellatum* var. *majus* · Forest floor plants · Ground disturbance · Rhizome type · Shoot elongation

6.1 Introduction

In mountain areas, slopes comprise several segments that are distinguished by changes in slope angle, which are termed as "breaks in slope" (Tamura 1969). Relatively active processes, such as soil erosion, landslides, and slope failures, occur more frequently on lower slope segments than on upper slopes and on ridge sites. Therefore, we can consider each segment as different habitats, which in turn have different types of vegetation established on it. In upper-stream mountain areas, the riparian forest corresponds to the forest on lower slope segments and on the valley bottom.

Generally, riparian forests have high species diversity. Herbaceous plants on the forest floor seem to largely contribute to the high species diversity (Kawanishi et al.

M. Kawanishi (✉)
Faculty of Education, Kagoshima University, Kagoshima, Japan
e-mail: kawanishi@edu.kagoshima-u.ac.jp

H. Sakio (ed.), *Long-Term Ecosystem Changes in Riparian Forests*, Ecological Research Monographs, https://doi.org/10.1007/978-981-15-3009-8_6

2008). The variation pattern of species diversity among habitats reflects differences in species coexistence patterns. Tree species distributions are generally limited by various combinations of disturbances and resources (Loehle 2000), but these have the potential for rapid migration (Clark 1998). On the other hand, the distribution patterns of forest floor herbaceous plants are determined by the availability of suitable habitats under the forest, the likelihood of seed dispersal to these habitats, and the successful germination of seeds and their subsequent growth (Ehrlen and Eriksson 2000; Gilliam and Roberts 2003). Thus, the effect of a disturbance differs for trees and for understory plants. This means that in order to clarify the species diversity pattern and mechanisms of the whole forest, we must first recognize the community structure independently for each life form (Kawanishi et al. 2008).

6.2 Comparison of Species Richness Among Different Deciduous Forest Corresponding to Slope Segment

To recognize patterns of species diversity, we compared the species richness of riparian forest among different deciduous forests based on their slope segments. We classified the mountain slope segments as crest slope, upper side slope, lower side slope, and valley-bottom, according to the hill slope system of Tamura (1987) (Fig. 6.1).

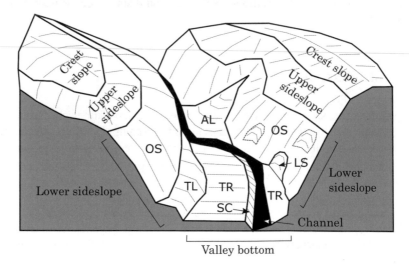

Fig. 6.1 Schematic diagram of landform types explained in this chapter (modified from diagram of Tamura 1987). Sub-small-scale landform types are shown as crest slope, upper side slope, lower side slope, and valley bottom, based on the hill slope system of Tamura (1987). Lower side slope and valley bottom are further sub-divided into micro-landform types (Kawanishi et al. 2004), i.e., terrace of debris flow (TR), alluvial fan (AL), terrace scarp (SC), new landslide site (LS), old landslide slope (OS), and talus (TL)

The valley vegetation of the Ooyamazawa river basin is a riparian forest that consists of *Fraxinus platypoda*, *Cercidiphyllum japonicum*, and *Pterocarya rhoifolia*, as was mentioned in a former chapter. On the other hand, *Fagus crenata–Fagus japonica* forests and *Tsuga sieboldii* forests are established on the upper side slopes and on the ridge, respectively (Maeda and Yoshioka 1952). Thus, species composition varies remarkably along the slope, and several forest types correspond to the micro-topography on the slope and in the watersheds (Kikuchi and Miura 1993; Sakai and Ohsawa 1994; Nagamatsu and Miura 1997). This structure of vegetation contributes in augmenting the plant species richness and diversity. In this chapter, we will discuss the distribution pattern of forest floor plants and how it relates to landforms, from the viewpoint of plant functional traits.

To recognize the pattern of species diversity, we attempted estimation using the hierarchical diversity model (Kawanishi et al. 2006). There are two levels, as follows: "*d*" is the sample quadrat diversity (Whittaker 1975), and "*D*" is the total diversity in a micro-landform unit. This hierarchical diversity model is a modified version of the model derived by Wagner et al. (2000).

The value of d is affected by within-quadrat species richness, but not by the quantitative dominance of species; instead, it is based on the species–area relationship (cf. formulae 6.1):

$$d = \frac{S}{\log_{10} A} \qquad (6.1)$$

where S is the number of species in a quadrat and A is the area of the quadrat. D (within-unit richness) shows the species richness per micro-landform, and was calculated as the total number of species in a micro-landform type (S_t) per total area (A_t, sum of the quadrat areas), such that:

$$D = \frac{S_t}{\log_{10} A_t} \qquad (6.2)$$

Figure 6.2 is a case study of vegetation in the Ooyamazawa river basin (Kawanishi et al. 2006). Species richness (d, D) was shown in each landform, and each Raunkiaer's life type (dormancy type) was classified based on the position of the dormant bud. This type represents the difference between the woody plants (MM, M, N, Ch), perennial herbs (H, G), and annual herbs (Th). Originally, this type was used to show the relationship between global climate and vegetation; however, we can also use this spectrum for overstory trees and forest floor plants.

On mountain slopes in the Ooyamazawa basin, indexes d and D of trees were higher on the upper side slope and crest slope than on the valley bottom and lower side slope (Fig. 6.2, see Fig. 6.1 for positional relation of landform). In contrast, the d of forest floor plants (Ch, H, G, Th) was very high on the valley bottom and lower side slope. These results indicate that the effects of topographical factors on species diversity differ between forest floor plants and overstory trees.

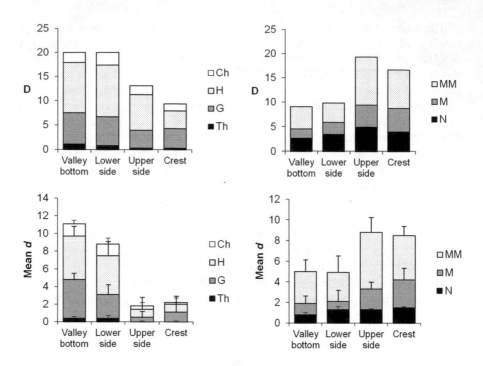

Fig. 6.2 Comparison of species richness for each life type among micro-landform types (Kawanishi et al. 2006). Indices *D* and means of index *d* are shown with standard deviations. Life types are Th: therophyte, G: geophyte, H: hemicryptophyte, Ch: chamaephyte, N: nanophanerophyte, M: microphanerophyte, and MM: megaphanerophyte

Why are overstory trees diverse in the upper slope areas? It would probably be related to the regeneration processes of trees, which depend on disturbances. For example, trees in beech forests on stable slopes generally regenerate in small canopy gaps when trees fall due to typhoons, etc. (Nakashizuka 1982, 1983, 1984). In addition, very few juvenile *Fagus crenata* are found in the beech forests on the Pacific Ocean side of Japan (including the Ooyamazawa river basin), although juveniles of many other species can be found (Shimano and Okitsu 1993, 1994). The regeneration processes in *Tsuga sieboldii* forests show similar patterns (Suzuki 1980). As a result, many small patches consisting of regenerate trees are allocated within a small area, producing a higher alpha diversity (d) for trees in the upper side slope and crest slope. In contrast, dominant trees on the valley bottom and lower side slope (e.g., *Pterocarya rhoifolia*, *Fraxinus platypoda*, and possibly *Cercidiphyllum japonicum*) generally regenerate simultaneously in the huge gaps created by rare, large disturbances (Sakio et al. 2002). Therefore, large disturbances would restrict the establishment of many deciduous trees that grow on the upper slope and would allow several trees to adapt to riparian disturbances. As a result, the index d for trees on the valley bottom and lower side slope is low. This tendency can be seen in other riparian forests, such as on the floodplain forest (Aruga et al. 1996) and on the relatively stable riparian terrace forest (Suzuki et al. 2002).

In contrast, the mean *d* of herbaceous plants on the valley bottom and on the lower side slope is very high. This indicates that frequent disturbances increase the diversity of forest floor plants. The reason for this tendency is the absence of strong competitors, e.g., dwarf bamboo (*Sasamorpha borealis*), which were removed by frequent ground disturbances in a riparian area. Because light reaching the forest floor is very scarce in dense dwarf bamboo communities (Nakashizuka 1988), other herbs may not be able to grow. In addition, we can observe the land heterogeneity of disturbance sites (Sakio 1997; Sakio et al. 2002). This heterogeneity will contribute to the high diversity of micro-habitat types relating to ground surface condition, such as gravel size, content ration of organic matter, and water content. Therefore, the various ground disturbances may be responsible for the high beta-diversity of herbs in riparian forests. This will be discussed in the next section.

6.3   Relationships Between Landforms and Life Type Composition in the Forest Floor Vegetation

The forest floor vegetation varies among different habitats based on their landform, as stated above. Given these findings, we sought characteristics of herbaceous plants that confer adaptation to various habitats. In general, the likelihood of seed dispersal to various habitats, and the successful germination of seeds and their subsequent growth determine the distribution of herbaceous plants (e.g., Ehrlen and Eriksson 2000; Gilliam and Roberts 2003). In this study, we focus on growth propagation, and we aim to show that the habitat restriction of herbaceous plants in riparian forests is caused by differences in breeding.

The characteristics of the life history of forest floor plants in Japan have been studied mainly in terms of reproductive ecology and seed ecology (e.g., Kawano 1975, 1985; Kawano and Nagai 1975). Generally, well-adapted to disturbances are annual herbs with short leaf lives and large growth amounts (Grime 2001). On the other hand, perennial herbs have various life history and life cycle characteristics that are related not only to environmental pressures or to interspecies competition, but also to their adaptation to disturbances (Kawano 1985). For example, some perennials have life history strategies equivalent to annual plants. Such plants are often called as "pseudo-annual plants". This includes *Cacalia delphiniifolia* (Fig. 6.3), *Cacalia tebakoensis*, *Senecio nikoensis*, and *Sanicula chinensis* (Numata and Asano 1969) which are interesting; however, there are still many unknown parts in their life history, so clarifying their significance is of great interest.

It is clear that the number of species capable of vegetative reproduction in the forest floor vegetation in the *Pterocarya rhoifolia* and *Fraxinus platypoda* forests is greater than that in the *Fagus crenata* and *Quercus crispula* forests (Oono 1996). Species that are early in making independent propagules from the mother individual

Fig. 6.3 Pseudo-annual
plants; *Cacalia
delphiniifolia*

by vegetative propagation are characteristically linked to the riparian *Fraxinus platypoda* forest. In general, the vegetative reproduction of herbaceous plants plays important roles in maintaining the population (Silvertown 1982; Grime 2001) and in recovering from damage caused by ground disturbance (Yano 1962). These indicate that the functional diversity of herbaceous plants greatly contributes to the species diversity of riparian vegetations. Therefore, I would like to describe the relationship between the life types of herbaceous plants and ground disturbances, with the aim of understanding the establishment of the forest floor vegetation in the riparian forests.

The habitat differentiation of herbs in the *Fraxinus playipoda* and *Pterocarya rhoifolia* forests is related to their vegetative reproduction characteristics (Kawanishi et al. 2004). In the Ooyamazawa basin, six landform types could be distinguished along the valley: debris flow terrace, alluvial fan, terrace scarp, new landslide site, old landslide slope, and talus (Fig. 6.1, Kawanishi et al. 2004). Forest floor plants were classified into 7 groups by cluster analysis, and we were able to identify three major groups (clusters A, B, D) (Table 6.1). Cluster A includes species belonging to spring ephemerals, storage rhizomes, and anti-vegetative reproduction. These species are perennial herbs and ferns with storage-type rhizomes, and they mainly grow on landforms such as debris flow terraces and alluvial fans where they are stable for long periods of time (Fig. 6.4, Kawanishi et al. 2004).

Spring ephemerals, such as *Corydalis lineariloba* (Fig. 6.5) and *Allium monanthum* (Fig. 6.6), have significantly greater concentrations of nitrogen and iron than other herbs (Muller 2003), which may relate to high anabolism. Storage organs have important roles in the effective distribution of carbohydrates and major

Table 6.1 Mean coverage (%) of species in each landform type with maximum coverage of species in parenthesis (Kawanishi et al. 2004, 2008)

Species name	Life type			Mean coverage (%)					
	GS	REP	MR	TR	AL	SC	LS	OS	TL
Cluster A									
Scopolia japonica	sm	lll	rs	8.1 (40)	7.4 (40)	2.1 (20)$^-$	1.6 (10)$^-$	3.2 (30)$^-$	9.5 (50)$^+$
Corydalis lineariloba	sp	tub	rr	7.0 (20)$^+$	9.0 (20)$^+$	4.3 (20)	1.7 (10)$^-$	3.6 (15)$^-$	3.0 (15)$^-$
Spuriopimpinella nikoensis	f	av	rs	1.5 (5)	5.5 (10)$^+$	2.2 (10)	0.5 (7)$^-$	0.6 (7)$^-$	1.4 (10)
Aconitum sanyoense	f	tub	rr	2.2 (15)$^+$	5.0 (30)$^+$	2.4 (20)$^+$	0.5 (10)$^-$	0.2 (3)$^-$	0.3 (7)$^-$
Veratrum grandiflorum	sm	rhi	rs	1.2 (10)	7.0 (20)$^+$	0.3 (5)$^-$	0.1 (2)$^-$	1.1 (20)$^-$	0.0 (0.1)$^-$
Mitella pauciflora	f	rhi	rc	4.5 (20)$^+$	2.4 (10)$^+$	0.4 (3)$^-$	0.7 (5)$^-$	0.4 (7)$^-$	0.4 (3)$^-$
Dryopteris crassirhizoma	f	av	rs	3.0 (10)$^+$	1.2 (10)	0.6 (5)$^-$	0.4 (5)$^-$	0.5 (10)$^-$	1.0 (10)
Cornopteris crenulato-serrulata	f	rhi	rc	2.4 (7)$^+$	1.9 (5)$^+$	0.3 (2)$^-$	0.5 (5)$^-$	0.3 (10)$^-$	0.4 (3)$^-$
Adoxa moschatellina	sm	rhi	rr	2.8 (10)$^+$	2.3 (15)$^+$	0.5 (2)$^-$	0.1 (2)$^-$	0.3 (5)$^-$	2.9 (15)$^+$
Allium monanthum	sp	tub	rr	3.1 (20)$^+$	0.5 (3)$^-$	0.7 (2)$^-$	0.7 (5)$^-$	0.7 (10)$^-$	1.3 (10)
Asarum caulescens	f	run	rc	2.4 (20)$^+$	0.9 (5)	0.9 (5)	0.3 (4)$^-$	1.3 (20)	0.2 (2)$^-$
Cacalia yatabei	f	rhi	rr	1.2 (15)$^+$	0.6 (2)	0.5 (7)	0.5 (10)	0.3 (3)$^-$	0.2 (3)$^-$
Polystichum tripteron	f	av	rs	1.4 (7)$^+$	0.1 (2)$^-$	0.7 (7)	0.4 (5)	0.0 (1)$^-$	0.8 (7)
Chrysosplenium ramosum	f	run	rr	1.0 (7)$^+$	0$^-$	0.4 (10)	0.4 (3)	0.3 (3)	0$^-$
Polystichum ovato-paleaceum	f	av	rs	0.4 (5)	0.1 (3)$^-$	0.6 (5)$^+$	0.8 (7)$^+$	0.2 (3)	0.1 (1)$^-$
Diplazium squamigerum	f	rhi	rc	1.6 (20)$^+$	0.2 (5)	0.2 (5)	0.0 (0.1)$^-$	0.3 (5)	0.2 (2)$^-$
Cluster B									
Chrysosplenium macrostemon	f	run	rr	0.6 (5)$^-$	0.1 (3)$^-$	1.0 (10)$^-$	4.5 (30)$^+$	3.8 (50)$^+$	0.1 (1)$^-$
Elatostema japonicum var. *majus*	f	bul/rhi	rc	0.2 (3)$^-$	0.0 (0.1)$^-$	3.6 (20)$^+$	5.2 (20)$^+$	2.0 (25)	0.0 (1)$^-$
Veronica miqueliana	f	rhi	rc	0.0 (1)$^-$	0.1 (2)$^-$	1.0 (7)$^+$	2.0 (10)$^+$	0.8 (7)	0.0 (0.1)$^-$
Persicaria debilis	f	th	rr	0.0 (0.1)$^-$	0$^-$	0.4 (7)	2.3 (10)$^+$	0.9 (10)$^+$	0.0 (0.1)$^-$
Cacalia farfaraefolia	f	rhi	rr	0.0 (1)$^-$	0.1 (1)$^-$	1.6 (7)$^+$	0.7 (3)	0.4 (5)	0.2 (5)$^-$
Laportea bulbifera	f	tub/bul	rr	0.8 (7)$^-$	0.5 (3)$^-$	1.3 (5)	2.6 (10)$^+$	1.4 (7)	2.5 (7)$^+$
Stellaria sessiliflora	f	run	rc	0.1 (3)$^-$	1.1 (20)	0.5 (7)	1.0 (7)	0.8 (5)	1.5 (5)$^+$
Impatiens noli-tangere	f	th	rr	0.6 (7)	0.0 (0.1)$^-$	0.7 (3)	1.2 (5)$^+$	0.4 (5)$^-$	1.7 (7)$^+$
Laportea macrostachya	f	rhi	rs	1.0 (10)	0.3 (3)$^-$	0.1 (2)$^-$	0.8 (3)	1.1 (7)	3.9 (20)$^+$
Deinanthe bifida	f	rhi	rs	0.2 (3)$^-$	0$^-$	0.3 (3)$^-$	2.1 (10)$^+$	2.5 (20)$^+$	0.4 (5)$^-$
Meehania urticifolia	f	run	rc	1.0 (5)	0.8 (3)	0.4 (3)$^-$	0.5 (5)	0.6 (5)	1.5 (7)$^+$
Dryopteris polylepis	f	av	rs	1.1 (7)$^+$	0.0 (0.1)$^-$	0.2 (3)$^-$	0.6 (3)	1.0 (5)$^+$	1.3 (5)$^+$
Cluster C									
Galium paradoxum	f	rhi	rr	0.2 (5)	0$^-$	0.1 (1)	0.8 (7)$^+$	0.0 (0.1)$^-$	0.8 (7)$^+$
Cluster D									
Hydrangea macrophylla var. *acuminata*	f	rhi	rc	0.2 (3)$^-$	0$^-$	3.0 (40)$^+$	1.3 (7)$^+$	0.4 (5)$^-$	0.5 (5)
Chrysosplenium album var. *stamineum*	f	run	rr	0$^-$	0$^-$	1.5 (7)$^+$	1.0 (7)$^+$	0.0 (2)$^-$	1.3 (20)$^+$
Astilbe thunbergii	f	rhi	rs	0$^-$	0$^-$	1.0 (10)$^+$	0.8 (10)$^+$	0.0 (0.1)$^-$	0.0 (0.1)$^-$
Cacalia delphiniifolia	f	rhi	rr	0.0 (1)$^-$	0.3 (7)	0.6 (7)$^+$	0.4 (5)	0.4 (3)	0.0 (0.1)$^-$
Cluster E									
Chrysosplenium echinus	f	run	rr	0.6 (7)$^+$	0.9 (10)$^+$	0.2 (2)$^-$	0.9 (15)$^+$	0$^-$	0.2 (3)
Cluster F									
Chrysosplenium pilosum var. *sphaerospermum*	f	run	rr	0.2 (7)$^-$	0.1 (3)$^-$	6.3 (30)$^+$	5.5 (25)$^+$	0.9 (15)$^-$	0.7 (15)$^-$
Cluster G									
Urtica laetevirens	f	rhi	rc	0.0 (1)$^-$	0.0 (0.1)$^-$	0.0 (0.1)$^-$	1.3 (20)$^+$	0.5 (15)	0.7 (10)$^+$

(continued)

Table 6.1 (continued) Superscript symbols "+" and "−" indicate desirable and undesirable site derived from χ^2 test ($P < 0.001$), respectively. *GS* growing season (sp: spring ephemeral, sm: summer period, f: three season), *REP* reproduction types (th: annual species, bul: bulbil type, tub: tuber type, run: runner type, rhi: horizontal rhizome type, and av: anti-vegetative reproduction type), and *MR* morphology of rhizome (rs: storage type, rc: connector type, and rr: replace type) are represented in column named "life type." See Fig. 6.1 for abbreviations of landform types (TR, AL, SC, LS, OS, TL)

nutrients in the plant body (e.g., Mooney and Billings 1961; Kimura 1970), and substances reserved in rhizomes sustain these species. Therefore, the distribution of these species would be restricted by breaks in the persistence of the storage organs. The rich organic matter in the soil of stable habitats, such as debris flow terraces and alluvial fans, may contribute to the maintenance of these plants (Fig. 6.4).

On the other hand, the other two species groups (B and D) characteristically comprised annual plants (such as *Impatiens noli-tangere*, Fig. 6.7 and *Persicaria debilis*, Fig. 6.8), perennials with bulbils (e.g., *Elatostema umbellatum* var. *majus.*, Fig. 6.9, *Laportea bulbifera*, Fig. 6.10), and plants with replacement rhizomes (e.g., *Chrysosplenium macrostemon*, Fig. 6.11, *Cacalia delphiniifolia* Sieb. et Zucc., Fig. 6.3, *Cacalia farfaraefolia* Sieb. et Zucc., means pseudo-annual). These species are dominant in locations where small annual disturbances occur frequently, like a sandbar along the stream and a new landslide site with unstable soils (Table 6.1). Generally, annual plants are adapted to unstable sites that experience continual or annual disturbances (Silvertown 1982). These aforementioned perennials would also be adapted to unstable habitats, because their life cycle is advantageous in maintaining populations that are subjected to soil disturbance, such as annual plants. These results indicate that the distribution pattern of herbaceous plants making up the forest floor vegetation is related to the attributes of their storage organs and to their vegetative reproduction properties.

6.4 How Do the Herbaceous Plants React to Micro-Ground Disturbance?

6.4.1 Three Different Perennial Plants

Whether or not a plant group can be maintained when the plant body is damaged by surface disturbance is expected to be related to the vegetative breeding style. Practically, how can the population of herb species be restricted by ground disturbance? There are new landslide sites in the foot part of the slope along the Ooyamazawa stream. In the newly collapsed site, the spring water from the pipe, which is thought to be the trigger of collapse, and the influence of the surface flow, which occurs at the time of rain because of steep inclination, are strong (Fig. 6.12). Therefore, the soil of the ground surface will move frequently over one year. Herbs growing in such a location must be largely influenced by how they can maintain their

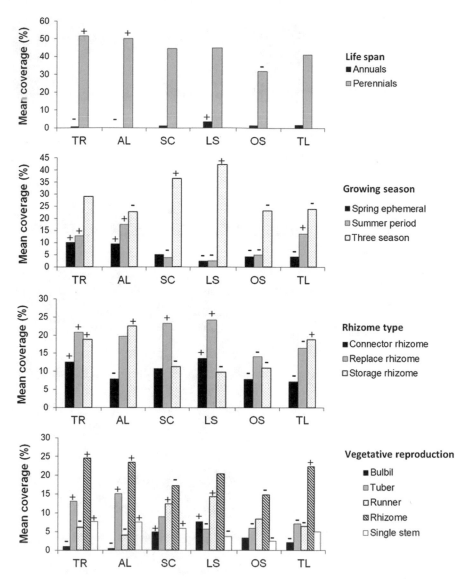

Fig. 6.4 Comparison of the mean coverage (%) of different life type groups in each geomorphic type (original data from Kawanishi et al. 2004). Mean coverages are shown and marked as desirable (+) or undesirable (−) sites, based on the χ^2 test ($P < 0.001$). See Fig. 6.1 for landform type abbreviations (TR, AL, SC, LS, OS, TL)

population, thanks to their resistance to this high frequency of disturbance. How does the life cycle of each species relate to disturbance?

Chrysosplenium macrostemon, *Elatostema japonicum* var. *majus*, and *Deinanthe bifida* are the major forest floor vegetation constituent species of the *Fraxinus playipoda* and *Pterocarya rhoifolia* forests, which are established on the Pacific

Fig. 6.5 *Corydalis lineariloba*

Fig. 6.6 *Allium monanthum*

side. These three species tend to appear on relatively unstable, small, collapsed terrains (Kawanishi et al. 2004); however, each has unique propagation characteristics.

Chrysosplenium macrostemon is a small perennial plant that breeds at the creeping stem on the ground surface (Fig. 6.13). The mother plant body dies after the new clone plant is formed at the apical bud and/or axillary bud of the creeping stem at the end of the growing season. This life cycle of *Chrysosplenium* resembles that of a pseudo-annual plant. On the other hand, *Elatostema japonicum var. majus* is a

Fig. 6.7 *Impatiens noli-tangere*

Fig. 6.8 *Persicaria debilis*

deciduous perennial that does not lose the main rhizome, but instead forms a bulbil in the node of the aerial stem and separates the vegetative propagation body every year (Fig. 6.14). On the other hand, *Deinanthe bifida* grows exclusively by underground stem: new underground shoots and old shoots are connected, and no vegetative propagation material to separate these is created (Fig. 6.15). Since these species are distributed in the most unstable collapsed places on the slope, this serves as a good reference to identify the relationship between propagation style and reaction to disturbance.

Fig. 6.9 *Elatostema*
umbellatum var. *majus*

Fig. 6.10 *Laportea*
bulbifera

From this point of view, we considered the mechanism of habitat selection of forest floor plants, focusing on the reactivity of individuals to disturbance. Firstly, we observed the fine-scale movement of the ground surface at the collapsed site. Secondly, leaf morphology, relating the shoot elongation and the reaction of the damaged individuals, was clarified. Finally, we considered the relationship between micro-disturbances and life types.

Fig. 6.11 Chrysosplenium
macrostemon

Fig. 6.12 The trace of small
piping phenomenon where
underground water has
spring out

6.4.2   Micro-disturbance on Small Landslide Site in Lower Side Slope

Because the new landslide site is the most unstable part of the slope, this site is suited for studying the tolerance of herbs to disturbances. So, we established an investigation plot on a small cliff part of the top and on the foot of a new landslide scar (Fig. 6.16). On this slope, there is clear knick line (convex break line) at the boundary of the upper valley side slope, and a small cliff, which is seen as a

Fig. 6.13 Whole plant
body of *Chrysosplenium
macrostemon*

Fig. 6.14 Whole plant
body of *Elatostema
japonicum* var. *majus*

Fig. 6.15 The flower and whole plant body of *Deinanthe bifida*

Fig. 6.16 Small landslide on the lower side slope

newly collapsed land in the new period that was made on the foot of slope. The survey plot was then set up in this newly collapsed site. *Chrysosplenium macrostemon*, *Elatostema japonicum* var. *majus*, and *Deinanthe bifida* are distributed in such unstable slopes, and involve slope failures among the riparian forests, as mentioned above. We can easily observe that the ramet of these herbs had damages, including breakage of stems, burial, dropout, etc. These may due to microdisturbances such as erosion of the surface accompanying piping as well as fine slippage.

6.4.3   Stem Elongation and Leaf Formation Pattern of Three Herbs

Elongation of the stem and the leaf formation in differently sized individuals are shown in Figs. 6.17, 6.18, and 6.19 for *Chrysosplenium macrostemon*, *Elatostema japonicum* var. *majus*, and *Deinanthe bifida*, respectively.

Figure 6.17 shows the shoot elongation and the leaves of *C. macrostemon* for small size (Cm-s), large size (Cm-l), and damaged (stem break) ramet (Cm-d). Regardless of the size of ramet, they begin extending the shoot from the overwinter rosette in early May, and they continue to grow until the beginning of September. Leaves (opposite phyllotaxis) were gradually attached at each node and corresponded with plant growth. For the large ramet, its side branches extended. The shoots stopped stem growth and leaf formation after winter rosette leaves formed on the shoot apex or on the lateral bud around late August. The rosette leaves of the previous year disappeared by the middle of June, and current leaves developed until the end of the observation at the end of October. With regard to the damaged ramet, though the stem had been broken in August, the stem tip of broken shoot continued to grow, formed foliage leaves, and subsequently formed overwinter rosette leaves. At the base of broken shoot, the elongation of the side branches improved slightly.

Shoot elongation and larvae of *D. bifida* are shown in Fig. 6.18 for the non-damaged small size (Db-s), medium size (Db-m), and the damaged large size shoot (Db-d). The schematic figure showing the rhizomes and the position of the above-ground shoots (Db-d) are also represented. The ground stems of *D. bifida* developed from the beginning of May and made two of three pairs of leaves until late June. On the other hand, the damaged shoot had relatively large dichotomous rhizomes (Fig. 6.18, Db-d), and the above-ground stems of this ramet started to grow from the tips of the thick rhizome in spring. Usually, although some preliminary buds had formed at the rhizomes of *D. bifida*, these buds do not elongate under safe conditions. However, when the above stem is damaged like that of shoot 1 (Fig. 6.18), preliminary sprouts begin to elongate at the branch of rhizome. Moreover, the preliminary sprouts (shoot 1-1) at the foot of damaged stem extended

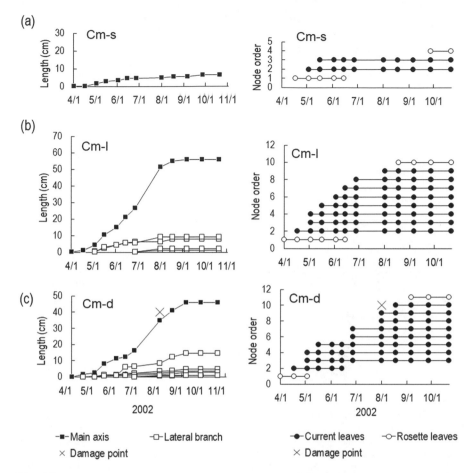

Fig. 6.17 Seasonal changes in the length of current shoots (left) and leaf survival states in each node on the main axis (right) of *Chrysosplenium macrostemon*. Small (Cm-s) and large (Cm-l) were shown as undamaged stems among growing seasons. The main axis of the shoot of the Cm-d individual was damaged (with a broken stem) in August. The survey was conducted in 2003

to about 5 cm past the damaged point. Other preliminary buds (shoot 2-1, 2-2) did not elongate.

Shoot elongation and leaf formation of E. *japonicum* var. *majus* are shown in Fig. 6.19. Non-damaged small (Ej-s), large (Ej-l) size shoot, and the damaged large size shoot (Ej-d) are shown as an example. Starting from early May, the ground stem of *E. japonicum* var. *majus* was growing alternate leaves while extending its shoot. Shoot elongation and leaf formation were almost finished in early August. The individuals shown in the figure formed a bulbil at the sixth node from the beginning of September. In damaged individuals (Ej-d), all stems had been buried by soil debris, making it impossible to accurately observe their reaction to the damage.

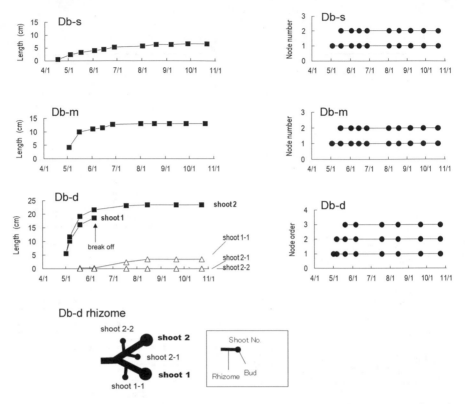

Fig. 6.18 Seasonal changes in the length of current shoots (left) and leaf survival states in each node on the main axis (right) of *Deinanthe bifida*. Small (Db-s) and medium (Db-m) size stems were shown as undamaged among growing seasons. The main stem of the shoot of the Db-d individual was damaged (stem was broken off) in June (Db-d). The survey was conducted in 2003. Rhizome pattern diagrams of damaged individuals (Db-d) are also shown. The survey was conducted in 2003

6.4.4 Adaptation to Micro-disturbance

In *C. macrostemon,* the shoots elongate until late summer, and as such, shoots can attach more leaves, which is similar to a long shoot. Shoots positively induce adventitious roots, after which stems will easily take root even if disturbances do not occur. In addition, the shoots could grow even if a stem had been broken or buried. These characteristics indicate that each shoot branch can be a ramet that is independent from the individual root even during the early seasons. Therefore, even if the plant body is damaged by ground disturbances, they can grow and form a vegetative propagation body, such as an overwintering rosette. Thus, it is possible to form vegetative propagules at the tip of each shoot and disperse the ramets every year.

The life history of *C. macrostemon* resembles pseudo-annual plants, as is described above. The characteristics of the habitats of pseudo-annuals in forests

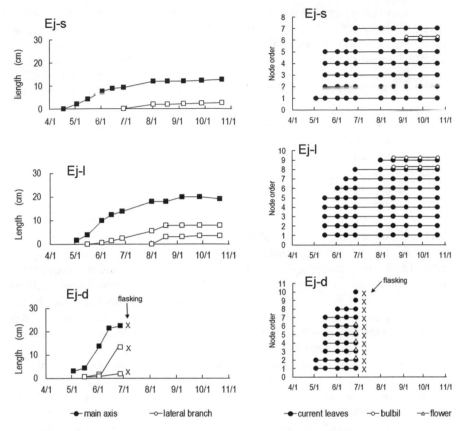

Fig. 6.19 Seasonal changes in the length of current shoots (left) and leaf survival states in each node of the main axis (right) of *Elatostema umbellatum* var. *majus*. Non-damaged small (Ej-s), large (Ej-l) size, and the damaged large size shoot (Ej-d) are shown. The survey was conducted in 2003

are that they are stable or somewhat unstable and experience no disturbance (Kawano 1985). In contrast, annuals, biennials, and perennials (making bulbil) tend to habit unstable sites with regular disturbances (Kawano 1985). This tendency is also shown in studies that compare different ecosystems (Svensson et al. 2013).

Coastal plants are examples of adaptation to permanent disturbance in unstable habitats. Coastal plants such as *Wedelia prostrata* with long horizontal stems (runner) and *Carex kobomugi* that extend long rhizomes underground are able to survive the strict beach environment (Yano 1962). This is because they can establish adventitious roots from the nodes on the runner or on the rhizome even if the above-ground shoots are buried by sand sedimentation or if the rhizomes are cut by wind erosion. This indicates the advantage of clonal plants in unstable sites. The life history of *C. macrostemon*, which involves generating adventitious roots while growing the creeping stems, may be adaptive on unstable slopes along the mountain stream where the ground surface moves finely.

Chrysosplenium plants grow mainly in unstable riparian areas along the mountain stream (Wakabayashi 2001), and several species distribute sympatrically in one region (Fukamachi et al. 2014). Fukamachi et al. (2014) clarified habitat environments and the overlap of the distribution of the *Chrysosplenium* species and pointed out that the micro-environment may contribute to coexistence. The mechanism of coexistence is not yet clear; thus, I would like to research the relationship between the shoot elongation pattern of each species and the slight variability of ground surface. On the other hand, shoot elongation and leaf formation of *D. bifida* start in spring, and are almost completed by early summer (Fig. 6.17), after which stems and leaves will no longer form. Because of this, the plant ends up having only 2–3 pairs of leaves. In general, the rhizome typically has a stouter stem than a stolon. Its old portion decays, separating the ramet into two new ramets when the rotting reaches a branch junction (Bell and Bryan 2008); *D. bifida* typically has these types of rhizome. The rhizome is relatively thick, stout, and has a lot of strong adventitious roots, with few that are branched. These growth patterns (i.e., simultaneous expansion of the leaf in spring) and morphology will indicate that they may primarily utilize the storage material of the rhizomes during new leaf formation. When shoots are damaged, preliminary sprouts of the rhizomes start to grow. Therefore, it is highly possible that the storage material of the rhizome is also needed in the growth of new preparative stems and leaves as recovery from the disturbance. Based on this fact, it is seen that *D. bifida* has an anabolic system that stores its annual assimilation products in rhizomes as much as possible.

E. umbellatum var. *majus* grows shoots until midsummer and gradually exhibits leaves. Both the shoot extension and the foliation are stopped and completed in the middle of August when the bulbil begins to form. From this time, the assimilation products seem to be also used for the formation of bulbils and storage of rhizomes. Unfortunately, we could not observe the reaction of damaged ramet was not clear. But we could observe the preliminary buds in the underground stem. Some preliminary sprouts may grow up from the bud in rhizomes if the upper shoot was damaged or lost.

And then, perennials having bulbil (such as *Laportea bulbifera*, *Sedum bulbiferum*, *Dioscorea bulbifera*, and *Lilium lancifolium*) grow on relatively unstable habitat (Kawano 1985). Bulbil of *E. umbellatum* var. *majus* also may have important role to maintain the population on unstable slope.

As described above, the elongation and development characteristics of shoots of these three species were different, and were thought to be closely related to the method of vegetative propagation. Reactivity to damage is determined by shoot growth, vegetative propagation characteristics, and how much storage of assimilation products has been done. In this book, we introduced only three species studied by the authors, but the life type of herbaceous plants constituting forest floor vegetation is diverse, and the life cycle and life history of most species are unknown. As introduced in Chap. 8, most of the current Japanese forests are affected by deer, and there are many areas where forest floor vegetation is declining. In order to examine its conservation and restoration, it is desirable to elucidate the mechanism by which species diversity of forest floor plants is maintained. For that purpose, we will need to advance more research on forest floor herbs.

References

Aruga M, Nakamura F, Kikuchi S, Yajima T (1996) Characteristics of floodplain forests and their site conditions in comparison to toeslope forests in the Tokachi River. J For Soc 78:354–362 (in Japanese with English abstract)

Bell A, Bryan A (2008) Plant form: an illustrated guide to flowering plant morphology (new edition). Timber Press, Inc., p 431

Clark JS (1998) Why trees migrate so fast: confronting theory with dispersal biology and paleorecord. Am Nat 152:204–224

Ehrlen J, Eriksson O (2000) Dispersal limitation and patch occupancy in forest herbs. Ecology 81:1667–1674

Fukamachi A, Hoshino Y, Ohashi H, Nakao K 2014 Distributional patterns and co-occurrence of Chrysosplenium species in watersheds in the upper Watarase river basin. Veg Sci 31:107–117 (in Japanese with English abstract)

Gilliam FS, Roberts MR (2003) Introduction, conceptual framework for studies of the herbaceous layer. In: Gilliam FS, Roberts MR (eds) The herbaceous layer in forest of eastern North America. Oxford University Press, New York, pp 3–11

Grime JP (2001) Plant strategies, vegetation processes, and ecosystem properties, 2nd edn. Wiley, Chichester

Kawanishi M, Sakio H, Ohno K (2004) Forest floor vegetation of *Fraxinus platypoda-Pterocarya rhoifolia* forest along Ooyamazawa valley in Chichibu, Kanto District, Japan, with a special reference to ground disturbance. Veg Sci 21:15–26 (in Japanese with English abstract)

Kawanishi M, Sakio H, Kubo M, Shimano K, Ohno K (2006) Effect of micro-landforms on forest vegetation differentiation and life-form diversity in the Chichibu Mountains, Kanto District, Japan. Veg Sci 23:13–24

Kawanishi M, Sakio H, Ohno K (2008) Diversity of forest floor vegetation with landform type. In: Sakio H, Tamura T (eds) Ecology of riparian forests in Japan: disturbance, life history, and regeneration. Springer, pp 267–278

Kawano S (1975) The productive and reproductive biology of flowering plants. II. The concept of life history strategy in plants. J Coll Liberal Arts Toyama Univ 8:51–86

Kawano S (1985) Life history characteristics of temperate woodland plants in Japan. In: White J (ed) The population structure of vegetation, handbook of vegetation science, vol 3. Dr W. Junk Publishers, Dordrecht, pp 515–549

Kawano S, Nagai Y (1975) The productive and reproductive biology of flowering plants - I. Life history strategies of three Allium species in Japan. Bot Mag Tokyo 88(4):281–318

Kikuchi T, Miura O (1993) Vegetation patterns in relation to micro-scale landforms in hilly land regions. Vegetatio 106:147–154

Kimura M (1970) Analysis of production processes of an undergrowth of subalpine Abies forest, Pteridophyllum racemosum population 1 Growth, carbohydrate economy and net production. Bot Mag Tokyo 83:99–108

Loehle C (2000) Strategy space and the disturbance spectrum: a life-history model for tree species coexistence. Am Nat 156:14–33

Maeda T, Yoshioka J (1952) Studies on the vegetation of Chichibu Mountain forest (2). The plant communities of the temperate mountain zone. Bull Tokyo Univ For 42:129–150 (in Japanese)

Mooney HA, Billings WD (1961) Comparative physiological ecology of arctic and alpine populations of Oxyria digyna. Ecol Monogr 31:1–29

Muller RN (2003) Nutrient relation of the herbaceous layer in deciduous forest ecosystems. In: Gilliam FS, Roberts MR (eds) The herbaceous layer in forest of eastern North America. Oxford University Press, New York, pp 15–37

Nagamatsu D, Miura O (1997) Soil disturbance regime in relation to micro-landforms and its effects on vegetation structure in a hilly area in Japan. Plant Ecol 133:191–200

Nakashizuka T (1982) Regeneration process of climax beech forest II. Structure of a forest under the influences of grazing. Jpn J Ecol 32:473–482

Nakashizuka T (1983) Regeneration process of climax beech forest III. Structure and development process of sapling populations in different aged gaps. Jpn J Ecol 33:409–418

Nakashizuka T (1984) Regeneration process of climax beech forest IV. Gap formation. Jpn J Ecol 34:75–85

Nakashizuka T (1988) Regeneration of beech (Fagus crenata) after the simultaneous death of undergrowing dwarf bamboo (Sasa kurilensis). Ecol Res 3:21–35

Numata M, Asano S (1969) Biological flora of Japan, vol 1. Tsukiji Shokan Publishing Co., Ltd. (in Japanese)

Oono K (1996) Life history of herb plants in summer green forest. In: Hara M (ed) Natural history of beech forest. Heibonsha, Tokyo, pp 113–156 (in Japanese)

Sakai A, Ohsawa M (1994) Topographical pattern of the forest vegetation on a river basin in a warm-temperate hilly region, central Japan. Ecol Res 9:269–280

Sakio H (1997) Effects of natural disturbance on the regeneration of riparian forests in Chichibu Mountains, central Japan. Plant Ecol 132:181–195

Sakio H, Kubo M, Shimano K, Ohno K (2002) Coexistence of three canopy tree species in a riparian forest in the Chichibu Mountains, central Japan. Folia Geobot 37:45–61

Shimano K, Okitsu S (1993) Regeneration of mixed Fagus crenata-Fagus japonica forests in Mt. Mito, Okutama, west of Tokyo. Jpn J Ecol 43:13–19 (in Japanese with English summary)

Shimano K, Okitsu S (1994) Regeneration of natural Fagus crenata forests around the Kanto district. Jpn J Ecol 44:283–291 (in Japanese with English summary)

Silvertown JW (1982) Introduction to plant population ecology. Longman, London, p 209

Suzuki E (1980) Regeneration of *Tsuga sieboldii* forest. II. Two cases of regenerations occurred about 260 and 50 years ago. Jpn J Ecol 30:333–346 (in Japanese with English summary)

Suzuki W, Osumi K, Masaki T, Takahashi K, Daimaru H, Hoshizaki K (2002) Disturbance regimes and community structures of a riparian and an adjacent terrace stand in the Kanumazawa Riparian Research Forest, northern Japan. Forest Ecol Manag 157:285–301

Svensson BM, Rydin H, Carlsson BÅ (2013) Clonality in the plant community. In: van der Maarel E, Franklin J (eds) Vegetation ecology, 2nd edn. Wiley-Blackwell, pp 141–163

Tamura T (1969) A series of micro-landform units composing valley-heads in the hills near Sendai. The science reports of the Tohoku University. 7th series, Geography 19(1):111–127

Tamura T (1987) Landform-soil features in the humid temperate hills. Pedologist 31:135–146 (in Japanese)

Wagner HH, Wildi O, Ewald KC (2000) Additive partitioning of plant species diversity in an agricultural mosaic landscape. Landsc Ecol 15:219–227

Wakabayashi M (2001) *Chrysosplenim* L. In: Iwatsuki K, Boufford DE, Ohba H (eds) Flora of Japan, vol IIb. Kodansha Ltd. Publishers, Tokyo, pp 58–70

Whittaker RH (1975) Communities and ecosystems, 2nd edn. Macmillan, New York

Yano N (1962) The subterranean organ of sand dune plants in Japan. J Sci Hiroshima Univ (Ser B, Div 2) 9:139–184 (in Japanese)

Chapter 7
Coexistence of Tree Canopy Species

Hitoshi Sakio and Masako Kubo

Abstract The canopy tree species *Fraxinus platypoda*, *Pterocarya rhoifolia*, and *Cercidiphyllum japonicum* coexist at the Ooyamazawa riparian forest research site. In this chapter, we clarify the coexistence mechanisms of riparian tree species as they pertain to disturbance regimes, life-history strategies, and responses to environmental factors. Reproductive strategies, e.g., seed production and germination, differ widely among these three species and we observed probable reproductive trade-offs in each species. Canopy-height individuals of *F. platypoda* are recruited from advanced saplings, and *P. rhoifolia* and *C. japonicum* both established following large-scale disturbance events. Basal sprouting, i.e., vegetative reproduction, is likely the mechanism by which *C. japonicum* survives and attains co-dominance in riparian forests. *F. platypoda* had greater shade and water tolerance than the other two species. Each of these species is well-adapted to the various disturbances typical of riparian zones. Therefore, the coexistence mechanisms among them are likely a combination of random chance and niche partitioning.

Keywords Coexistence · Disturbance regime · Germination · Life history · Reproductive strategy · Seed production · Seedling survival · Shade tolerance · Vegetative reproduction · Water tolerance

H. Sakio (✉)
Sado Island Center for Ecological Sustainability, Niigata University, Niigata, Japan
e-mail: sakio@agr.niigata-u.ac.jp; sakiohit@gmail.com

M. Kubo
Faculty of Life and Environmental Science, Shimane University, Shimane, Japan
e-mail: kubom@life.shimane-u.ac.jp

© The Author(s) 2020

H. Sakio (ed.), *Long-Term Ecosystem Changes in Riparian Forests*, Ecological Research Monographs, https://doi.org/10.1007/978-981-15-3009-8_7

7.1 Introduction

Natural disturbances and life-history characteristics are key factors influencing the coexistence of tree species (White 1979; Loehle 2000). Disturbances in riparian areas are dynamic and vary more widely in type, frequency, and magnitude compared with those on hillsides. Various disturbance regimes in riparian areas lead to heterogeneous topography (Gregory et al. 1991; Kovalchik and Chitwood 1990) due to repeated destruction and regeneration of riparian forests.

Fraxinus platypoda, *Pterocarya rhoifolia*, and *Cercidiphyllum japonicum* (Fig. 7.1) are the dominant tree species in the riparian forests of the Chichibu

Fig. 7.1 The three dominant canopy tree species in the Ooyamazawa riparian forest of the Chichibu Mountains in central Japan

Mountains in central Japan (Maeda and Yoshioka 1952; Sakio 1997). One such riparian forest, along the Ooyamazawa stream, is an old-growth forest whose high-elevation trees have not been affected by human impacts, e.g., logging or erosion control works (Chap. 1), and are therefore valuable. The dominant canopy tree species in this forest is *F. platypoda*, followed by *P. rhoifolia* and *C. japonicum*.

The life-history characteristics and regeneration processes of these three canopy tree species were explained in detail in Chaps. 2, 3 and 4. In this chapter, we clarify the coexistence mechanism of riparian tree species in terms of disturbance regimes, tree life-history strategies, and responses to environment factors.

7.2   Seed Production

All three dominant canopy species in the Ooyamazawa riparian forest produce wind-dispersed winged achenes (Fig. 7.2). The average dry weights (mean ± standard deviation [SD], $n = 20$) of the fruit of *F. platypoda*, *P. rhoifolia*, and *C. japonicum* are 144 ± 24, 90 ± 11, and 0.82 ± 0.15 mg, respectively. The dry weights (mean ± SD, $n = 20$) of their seeds are 80 ± 17, 70 ± 8, and 0.58 ± 0.14 mg, respectively (Sakio et al. 2002).

The seeds of these three species are released according to different schedules, from autumn to winter (Fig. 7.3).The seeds of *P. rhoifolia* are released in October, whereas those of *F. platypoda* are released in November, during leaf fall. *C. japonicum* seeds are released after leaf fall, from November until spring of the

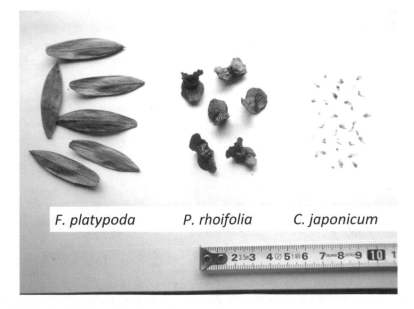

F. platypoda   *P. rhoifolia*   *C. japonicum*

Fig. 7.2 Seeds of the three riparian species

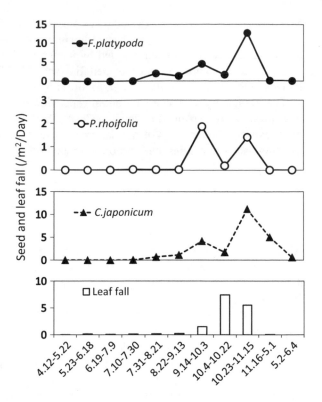

Fig. 7.3 Seed dispersal of the three riparian species and leaf fall in 2002. Leaf fall includes all tree species

following year. The dispersal distances of *P. rhoifolia* and *F. platypoda* are similar, extending to several tens of meters. In contrast, the significantly lighter seeds of *C. japonicum* (Welch's *t*-test, $P < 0.001$) are dispersed over hundreds of meters, with the maximum dispersal distance recorded being 302 m (Sato et al. 2006).

Secondary dispersal of *F. platypoda* seeds can also occur via water, implying an adaptation for waterborne fruit dispersal. We have observed many *F. platypoda* seeds being transported by the flow of mountain streams during mast seed years (Chap. 2); this process allows the establishment of many seedlings on the gravel banks of mountain streams in the following year (Fig. 7.4).

Annual fluctuation in seed production varies among tree species. Generally, late successional species have large seeds and irregular fruiting behavior, whereas pioneer species have small seeds and regular fruiting. *F. platypoda* had non-mast years in 1997, 2001, 2003, 2005, and 2015, and mast years in 1996, 1998, 2002, 2004, 2006, and 2016 (Fig. 7.5). Thus, this species exhibits irregular fruiting behavior, as has also been observed in *Fraxinus excelsior*, which is native throughout mainland Europe (Tapper 1992, 1996). Seed production also fluctuates annually in *P. rhoifolia*, which had non-mast years in 1996, 2006, and 2008, and mast years in 2005, 2007, and 2011. In contrast, *C. japonicum* produces regular amounts of seeds every year, with only slight fluctuation. Among these three species, *F. platypoda* had

Fig. 7.4 Current-year
F. platypoda seedlings
germinated at the stream
edge. Some seedlings are
submerged

the largest coefficient of variation (CV) of seed production (0.97), followed by
P. rhoifolia (0.87) and *C. japonicum* (0.48).

7.3   Seedling Germination and Growth

7.3.1   Germination Sites

Seeds of the three species germinate from mid-May to early summer of the year
following seed production. Buried seeds in Ooyamazawa riparian forest soil germinated no *F. platypoda* seeds, one *P. rhoifolia* seed, and 12 *C. japonicum* seeds in
30 L of soil (Kubo et al. 2008). Seeds of *P. rhoifolia* and *C. japonicum* have
exhibited dormancy in nursery seedling tests; in particular, *P. rhoifolia* seeds have
successfully germinated 2 years after sowing.

Germination sites differ among the three species (Fig. 7.6). Figure 7.6 shows the
results of a survey conducted in a year following a poor year for *F. platypoda* seed
production (all seedlings aged >1 year). But, in the following year, a mast year,
F. platypoda seeds germinated in all environments. *F. platypoda* can germinate in
forest floor environments varying in light, substrate, and water conditions, including
at water edges, under herb cover, and on steep slopes and gravel. Most *F. platypoda*
seedlings that germinate in the litter layer die due to fungal damage and lack of
moisture; however, some individuals survive for several years (Chap. 2). Current-
year seedlings have a very high survival rate on the gravel banks of mountain
streams, where herbaceous vegetation is rare. *F. platypoda* seedlings are highly
shade-tolerant and tend to concentrate in former stream channels and small gravel
deposits (Sakio 1997). When a canopy gap forms, thus improving the light environment, *F. platypoda* seedlings begin to grow into canopy trees.

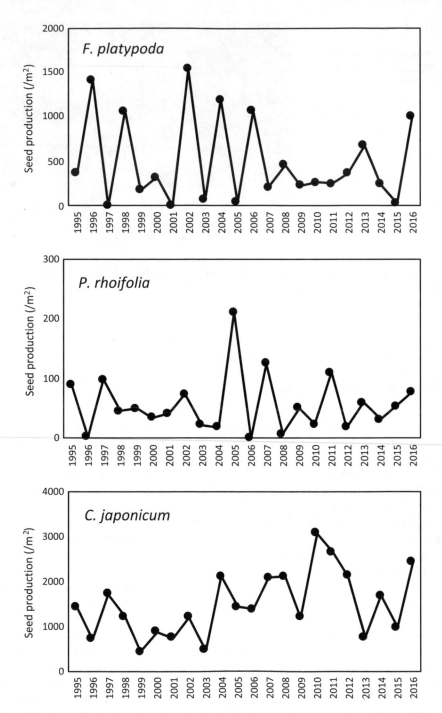

Fig. 7.5 Annual fluctuation of seed production in the three riparian species from 1995 to 2016

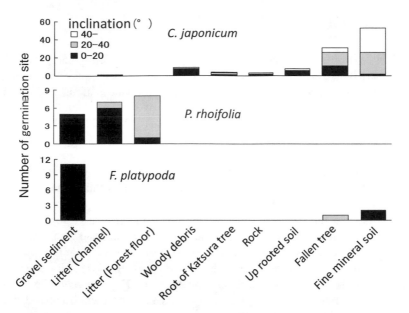

Fig. 7.6 Seed germination sites of the three riparian species (Kubo et al. 2000)

P. rhoifolia can also germinate in any environment, except on steep slopes due to unstable substrates. *P. rhoifolia* germination has been observed under closed canopies, and in gravel deposits and the litter layer; however, seedlings that germinate in the litter layer die within 1–2 months of germination (Sakio et al. 2002). Surviving individuals develop true leaves, but their growth is greatly affected by light availability. *P. rhoifolia* seedlings are affected by herbal pressure, even in canopy gaps; they require a brighter environment for survival than *F. platypoda* and may show arrested tree growth in low-light environments (Kisanuki et al. 1995).

In contrast, *C. japonicum* seeds germinate in a limited range of environments. Most *C. japonicum* seedlings have been found in fine mineral soil and on fallen logs. Seedling emergence of small-seeded species is generally reduced in litter deposits (Seiwa and Kikuzawa 1996); the germination and establishment of small-seeded species, such as birch (*Betula platyphylla* var. *japonica*), is promoted in fine-grained soil (Koyama 1998). These factors also appear to affect *C. japonicum* seed germination. Germination begins in mid-May; however, seedlings can be washed away by surface sediment movement during the rainy and typhoon seasons, with a survival rate of <10% in late October. This rate is higher at sites with high illuminance, and seedlings found in such environments have been reported to be significantly larger than those in low-light environments (Kubo et al. 2000). Similar results were obtained in a germination experiment examining the effects of soil and relative photosynthetic effective photon flux density on germination in a nursery (Kubo et al. 2004).

7.3.2 Seedling Shade Tolerance

The light environment is among the most important factors for plant growth. Plant responses to light differ among plant species. Pioneer species require more light than late successional species. In forests, canopy gap formation strongly affects light conditions. Improvement of the light environment by gap formation is an important factor for growth from seedling to canopy tree (Suzuki 1980, 1981; Nakashizuka and Numata 1982a, b; Nakashizuka 1983, 1984). For seedlings that have established under the canopy, a lack of canopy gap formation within a few years will result in decreased growth and, eventually, death. If an individual establishes under a small canopy gap, and that gap closes due to branch extension within the surrounding canopy, the individual will be unable to survive.

We compared the effects of light conditions on the growth of seedlings of the three Ooyamazawa riparian species in a nursery. *P. rhoifolia* had the fastest growth rate, growing to 30-cm seedlings within 1 year, whereas *F. platypoda* and *C. japonicum* seedlings measured about 10 cm (Fig. 7.7). However, since *C. japonicum* seeds are much smaller than those of *F. platypoda* (Fig. 7.2), its relative growth rate was very high. *P. rhoifolia* seedling growth decreased sharply under nursery light conditions (<20% of outdoor sunlight), and failed to survive in 1% light. Similarly, *C. japonicum* seedlings did not survive at 1% light. In contrast, all *F. platypoda* seedlings survived for 1 year at 1% light (Fig. 7.7; Sakio 2008). A study conducted at the Ooyamazawa riparian forest research site showed that branch growth was dramatically faster in 1-m-tall *P. rhoifolia* saplings than in *F. platypoda* saplings of the same height beneath a canopy gap (Sakio 1993). At a seedling size of <1 m, *F. platypoda* grew in lower light environments than *P. rhoifolia* and *C. japonicum* at the same research site (Kubo et al. 2000).

Fig. 7.7 Relationship between growth and relative light intensity for current-year seedlings grown in nursery beds for 1 year (Sakio 2008)

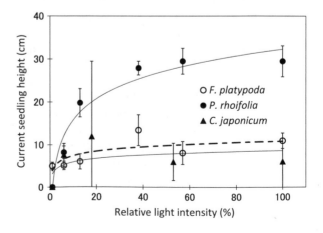

7.3.3 Seedling Water Tolerance

Trees in riparian areas are always exposed to a moist environment at the water's edge, and seedlings that have germinated on the gravel banks of mountain streams are affected by flooding during the rainy and typhoon seasons (Fig. 7.4). Water tolerance thresholds differ greatly among the three riparian species. Submergence experiments, in which 1-year-old seedlings were submerged to the soil surface for 1 year (Sakio 2005) and current-year seedlings were submerged for different periods (Fig. 7.8; Sakio 2008), showed that *F. platypoda* has much higher water tolerance

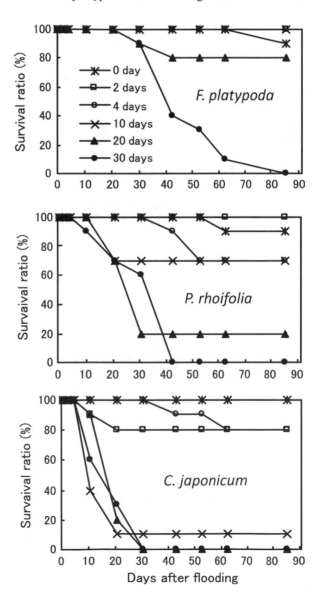

Fig. 7.8 Flooding survival rates of current-year seedlings of the riparian species

than do *P. rhoifolia* and *C. japonicum*. Dry weight was significantly lower in submerged 1-year-old *P. rhoifolia* and *C. japonicum* seedlings than in control individuals; moreover, 80% of *F. platypoda* current-year seedlings survived 20 days of submergence, whereas only 20% of *P. rhoifolia* seedlings survived, and all *C. japonicum* seedlings died. The superior water tolerance of *F. platypoda* may be one reason explaining its dominance in Ooyamazawa riparian forests.

7.4   Sprouting

All three riparian tree species exhibit reproduction by sprouting, with most *F. platypoda* producing single trees (Fig. 7.9; Sakio et al. 2002). *P. rhoifolia* produces intermediate numbers of sprouts; *C. japonicum* produces the most sprouts, with a maximum of 60 observed from one tree. In *C. japonicum*, sprout number was positively correlated with the diameter at breast height (DBH) of the main stem. The role of the *C. japonicum* sprouts is maintenance of the individual; this has also been observed in *Euptelea polyandra*, which maintains individuals by sprouting in areas frequently disturbed by landslides (Sakai et al. 1995). However, it remains unknown whether the sprouting mechanism of *C. japonicum* is related to physical damage to the individual, physiological responses to changes in the light environment, or tree age. In *P. rhoifolia*, sprouting does not play a role in maintenance of the individual in Ooyamazawa riparian forests, although such maintenance by sprouting has been observed in *P. rhoifolia* growing in environments characterized by heavy snow (Nakano and Sakio 2017, 2018).

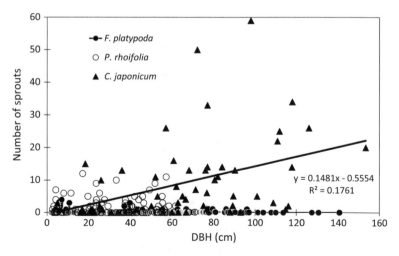

Fig. 7.9 Relationship between the diameter at breast height (DBH) of the main stem and sprout number for each individual of the three riparian species (Sakio 2008)

7.5 Forest Structure

7.5.1 Size Structure

A survey of all living trees (DBH ≥ 4 cm) without sprouts in an area of 4 71 ha showed that the frequency distribution of DBH was similar between *F. platypoda* and *P. rhoifolia*, being characterized by many small individuals with DBH ≤ 10 cm or less (Fig. 7.10). *C. japonicum* exhibited a different DBH distribution, with only two saplings at DBH < 10 cm. Both *F. platypoda* and *P. rhoifolia* showed frequency distribution peaks around DBH = 40. Within a core plot (0.54 ha), there were 811 *F. platypoda*, 192 *P. rhoifolia*, and one *C. japonicum* individual with tree height ≥ 1 m and DBH < 4 cm. Many *F. platypoda* and *P. rhoifolia* seedlings with DBH < 4 cm were distributed beneath the canopy (Chaps. 2 and 3). These results demonstrate that *F. platypoda* and *P. rhoifolia* produce many advanced seedlings that will eventually become canopy trees.

7.5.2 Spatial Distribution and Age Structure

Figure 7.11 shows the spatial distribution of canopy trees (DBH ≥ 20 cm) and young trees (4 cm ≤ DBH < 20 cm) within the 4.71-ha research plot. Canopy and young trees of *F. platypoda* are dominant from upstream to downstream, whereas *P. rhoifolia* canopy trees are distributed in three patches (A, B, and C). Large patches of canopy trees can reach 50 m in diameter, each with an average tree age of about 90 years. These *P. rhoifolia* patches were topographically distributed on

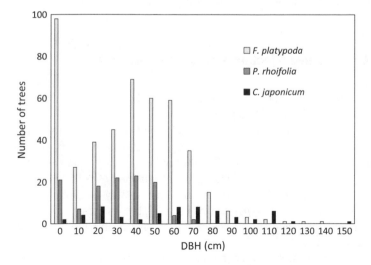

Fig. 7.10 DBH class distributions of the three riparian species (Sakio 2008)

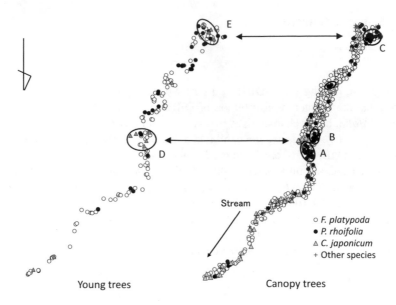

Fig. 7.11 Spatial distribution of young and canopy trees, including at the sites of two large-scale disturbances. *P. rhoifolia* canopy trees and young *C. japonicum* trees coexist at the same sites

large landslide or debris flow paths, suggesting that *P. rhoifolia* established in these large-scale disturbance sites at the same time (Sakio et al. 2002).

C. *japonicum* canopy trees are few in number and scattered randomly throughout the research area, although young trees showed aggregated distributions in and around *P. rhoifolia* patches (Fig. 7.11, D and E). At approximately 90 years, the ages of young *C. japonicum* and *P. rhoifolia* trees were similar; therefore, these two species are likely to have established at the same time.

7.6   Coexistence Mechanisms of the Three Species

The following coexistence mechanisms of canopy species in the Ooyamazawa riparian forest can be considered (Fig. 7.12).

In the Ooyamazawa riparian forest, *F. platypoda* regenerated to produce canopy trees through seedling establishment on large-scale disturbance sites, such as landslide and debris flow paths, as well as through the release of advanced forest floor seedlings upon the formation of small gaps. *F. platypoda* has higher shade tolerance than *P. rhoifolia* and *C. japonicum*, as demonstrated by seedling field studies and nursery experiments. *F. platypoda* also shows higher tolerance to submergence and flooding than *P. rhoifolia* and *C. japonicum*, and has adapted to germinate and grow on gravel stream banks.

The 4.71-ha research plot contained several large patches of *P. rhoifolia* with DBH of about 50 cm; these were established on the sites of large-scale disturbances

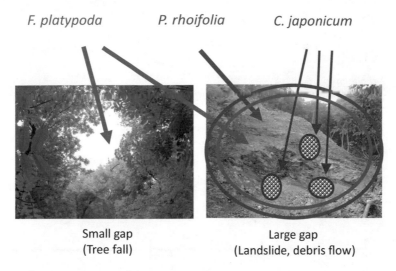

F. platypoda   *P. rhoifolia*   *C. japonicum*

Small gap   Large gap
(Tree fall)   (Landslide, debris flow)

Fig. 7.12 Mechanism for the coexistence of three canopy tree species in the Ooyamazawa riparian area

that occurred about 90 years ago. Thus, *P. rhoifolia* regeneration sites are limited to large-scale disturbance sites where large gaps are formed by landslide or debris flow; the regenerants eventually became canopy trees, forming single-species patches.

Unlike *F. platypoda*, *C. japonicum* cannot form large cohorts of advanced seedlings on the forest floor beneath canopy trees. However, *C. japonicum* seeds can disperse into the large-scale disturbance sites favored by *P. rhoifolia*. The mass production of small wind-dispersed seeds increases the probability of reaching new germination sites cleared by large disturbances (Harper 1977; Augspurger 1984). Thus, *C. japonicum* regeneration in large-scale disturbance sites can co-occur with that of *P. rhoifolia*. In these disturbance sites, organic and inorganic matter such as fallen trees, boulders, soil, and sand are mixed together to form complex substrates. The diversity of microsites produced during this process ensures seed germination, seedling establishment, and growth of *C. japonicum*. Fine inorganic soils, which form in boulders and the cracks of fallen trees, are ideal substrates for *C. japonicum* germination and establishment. These substrates are less susceptible to erosion due to rainfall, such that seedlings can grow stably; if direct sunlight is weak, seedlings may continue to grow because strong light dries out these soils. Thus, *C. japonicum* waits for rare regeneration opportunities, which appear at the scale of decades or centuries; as a trade-off, the lifespan of the individual is prolonged. *C. japonicum* generates many sprouts around the main trunk. In *F. platypoda* and *P. rhoifolia*, the death of the trunk means the inevitable death of the individual; however, if the main trunk of *C. japonicum* dies, one of the many surrounding sprouts will grow to become the main trunk (Fig. 7.13). In this manner, once *C. japonicum* has established, it will survive for long periods by successively producing new trunks.

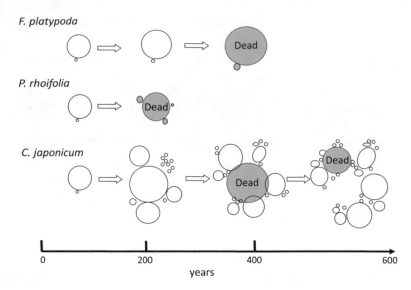

Fig. 7.13 Relationship between tree age and sprouting. In *F. platypoda* and *P. rhoifolia*, the death of the main trunk results in the death of the individual. In contrast, *C. japonicum* individuals are maintained by sprouting for long periods after the death of the main trunk

Therefore, *C. japonicum* trees represent a small population within the Ooyamazawa riparian forest, but this species sustainably coexists with both the shade-tolerant dominant species *F. platypoda* and the pioneer species *P. rhoifolia*.

7.7   Conclusion

The coexistence mechanisms of the three Ooyamazawa riparian canopy species involve a combination of niche partitioning and chance. The three species exhibit trade-offs in reproductive characteristics, e.g., seed size, quantity, and annual variation. Coexistence is generally maintained through niche partitioning, especially in the early life-history stages. Chance can also play an important role in *P. rhoifolia* and *C. japonicum* regeneration, through unpredictable large-scale, low-frequency disturbance. In conclusion, *F. platypoda*, *P. rhoifolia*, and *C. japonicum* are well-adapted to disturbances in the Ooyamazawa riparian zone throughout their life histories.

References

Augspurger CK (1984) Seedling survival of tropical tree species: interactions of dispersal distance, light gaps, and pathogens. Ecology 65(6):1705–1712
Gregory SV, Swanson FJ, Mckee WA, Cummins KW (1991) An ecosystem perspective of riparian zones: focus on links between land and water. BioScience 41:540–551

Harper JL (1977) Population biology of plants. Academic Press, London

Kisanuki H, Kaji M, Suzuki K (1995) The survival process of ash (*Fraxinus spaethiana* Ling.) and wingnut (*Pterocarya rhoifolia* Sieb. et Zucc.) seedlings at the riparian forest at Chichibu Mountains. Bull Tokyo Univ For 93:49–57 (in Japanese with English summary)

Kovalchik BL, Chitwood LA (1990) Use of geomorphology in the classification of riparian plant associations in mountainous landscapes of central Oregon, USA. Forest Ecol Manag 33/34:405–410

Koyama H (1998) Germination strategy of *Betura platyphylla* var. *japonica* (IV). advantage of small seeds for desiccation tolerance. Northern Forestry, Japan 50:276–280 (in Japanese)

Kubo M, Shimano K, Sakio H, Ohno K (2000) Germination sites and establishment conditions of *Cercidiphyllum japonicum* seedlings in the riparian forest. J Jpn For Soc 82:349–354 (in Japanese with English summary)

Kubo M, Sakio H, Shimano K, Ohno K (2004) Factors influencing seedling emergence and survival in *Cercidiphyllum japonicum*. Folia Geobot 39:225–234

Kubo M, Kawanishi M, Shimano K, Sakio H, Ohno K (2008) The species composition of soil seed banks in the Ooyamazawa riparian forest, in the Chichibu Mountains, central Japan. J Jpn For Soc 90:121–124 (in Japanese with English summary)

Loehle C (2000) Strategy space and the disturbance spectrum: a life-history model for tree species coexistence. Am Nat 156:14–33

Maeda T, Yoshioka J (1952) Studies on the vegetation of Chichibu Mountain forest (2). The plant communities of the temperate mountain zone. Bull Tokyo Univ For 42:129–150 (in Japanese with English summary)

Nakano Y, Sakio H (2017) Adaptive plasticity in the life history strategy of a canopy tree species, *Pterocarya rhoifolia*, along a gradient of maximum snow depth. Plant Ecol 218(4):395–406

Nakano Y, Sakio H (2018) The regeneration mechanisms of a *Pterocarya rhoifolia* population in a heavy snowfall region of Japan. Plant Ecol 219(12):1387–1398

Nakashizuka T (1983) Regeneration process of climax beech (*Fagus crenata* Blume) forests. III. Structure and development processes of sapling populations in different age gaps. Jpn J Ecol 33:409–418

Nakashizuka T (1984) Regeneration process of climax beech (*Fagus crenata* Blume) forests IV. Gap formation. Jpn J Ecol 34:75–85

Nakashizuka T, Numata M (1982a) Regeneration process of climax beech forests I. Structure of a beech forest with the undergrowth of *Sasa*. Jpn J Ecol 32:57–67

Nakashizuka T, Numata M (1982b) Regeneration process of climax beech forests II. Structure of a forest under the influences of grazing. Jpn J Ecol 32:473–482

Sakai A, Ohsawa T, Ohsawa M (1995) Adaptive significance of sprouting of *Euptelea polyandra*, a deciduous tree growing on steep slope with shallow soil. J Plant Res 108:377–386

Sakio H (1993) Sapling growth patterns in *Fraxinus platypoda* and *Pterocarya rhoifolia*. Jpn J Ecol 43(3):163–167 (in Japanese with English summary)

Sakio H (1997) Effects of natural disturbance on the regeneration of riparian forests in a Chichibu Mountains, central Japan. Plant Ecol 132:181–195

Sakio H (2005) Effects of flooding on growth of seedlings of woody riparian species. J Forest Res 10:341–346

Sakio H (2008) Coexistence mechanisms of three riparian species in the upper basin with respect to their life histories, ecophysiology, and disturbance regimes. In: Sakio H, Tamura T (eds) Ecology of riparian forests in Japan: disturbance, life history and regeneration. Springer, pp 75–90

Sakio H, Kubo M, Shimano K, Ohno K (2002) Coexistence of three canopy tree species in a riparian forest in the Chichibu Mountains, central Japan. Folia Geobot 37:45–61

Sato T, Isagi Y, Sakio H, Osumi K, Goto S (2006) Effect of gene flow on spatial genetic structure in the riparian canopy tree *Cercidiphyllum japonicum* revealed by microsatellite analysis. Heredity 96:79–84

Seiwa K, Kikuzawa K (1996) Importance of seed size for the establishment of seedlings of five deciduous broad-leaved tree species. Vegetatio 123:51–64

Suzuki E (1980) Regeneration of *Tsuga sieboldii* forest. II. Two cases of regenerations occurred about 260 and 50 years ago. Jpn J Ecol 30:333–346 (in Japanese with English synopsis)

Suzuki E (1981) Regeneration of *Tsuga sieboldii* forest. III. Regeneration under a canopy gap with low density level of conifer seedlings and a method for estimating the time of gap formation. Jpn J Ecol 31:307–316 (in Japanese with English synopsis)

Tapper PG (1992) Irregular fruiting in *Fraxinus excelsior*. J Veg Sci 3:41–46

Tapper PG (1996) Long-term patterns of mast fruiting in *Fraxinus excelsior*. Ecology 77:2567–2572

White PS (1979) Pattern, process and natural disturbance in vegetation. Bot Rev 45:229–299

Part IV
Ecosystem Changes in Riparian Forests

Chapter 8
Changes in Vegetation in the Ooyamazawa Riparian Forest

Forest Floor Vegetation

Motohiro Kawanishi, Masako Kubo, Motoki Higa, and Hitoshi Sakio

Abstract Sika deer are responsible for diverse effects in the forest vegetation of the Chichibu Mountains. The lowest level of understory plants in the riparian forest of Ooyamazawa in the Chichibu Mountains covered over 90% of the land area in 1983, but the coverage decreased to 3% by 2004. In 2017, the area seemed to have recovered slightly; the coverage was approximately 10%. The decreases in plant populations could be attributed to sudden increases in deer population density, detected post-2000 by cervid research in this area. The plants that remained as of 2017 shared characteristics which could have contributed to their survival in the presence of deer overpopulation. They were either I: toxic plants, II: small herb species which do not increase in plant height, or III: perennial herbs in which the aerial portion withers until the summer and enters yearly dormancy at an early date.

Keywords Deer fence · Ferns · Forest floor vegetation · Grazing · Herbs · Sika deer · Species diversity · Toxic plants · Understory plants

M. Kawanishi (✉)
Faculty of Education, Kagoshima University, Kagoshima, Japan
e-mail: kawanishi@edu.kagoshima-u.ac.jp

M. Kubo
Faculty of Life and Environmental Science, Shimane University, Matsue, Japan
e-mail: kubom@life.shimane-u.ac.jp

M. Higa
Faculty of Science and Technology, Kochi University, Kochi, Japan
e-mail: mhiga@kochi-u.ac.jp

H. Sakio
Sado Island Center for Ecological Sustainability, Niigata University, Niigata, Japan
e-mail: sakio@agr.niigata-u.ac.jp; sakiohit@gmail.com

© The Author(s) 2020
H. Sakio (ed.), *Long-Term Ecosystem Changes in Riparian Forests*, Ecological Research Monographs, https://doi.org/10.1007/978-981-15-3009-8_8

139

8.1   Introduction

In recent years, the number of sika deer (*Cervus nippon*) broadly distributed throughout the Shikoku, Kyushu, and Honshu districts in Japan has increased, and their negative impact on various forest vegetation is evident (e.g., Hattori et al.2010; Ishida et al.2012; Otsu et al.2011; Shimoyama 2012; Planning Committee-The Society of Vegetation Science 2011; Ohashi et al.2014). The influence of deer grazing in forests appears in the forest floor vegetation at an early stage; unpalatable plants begin to dominate as the grazing effects intensify (Takatsuki 2009).

A long-term monitoring study that began before the vegetation damage increased in Kanto district (Okutama region) by Ohashi et al. (2007) showed that the height and coverage of the herbaceous layer had tended to decrease in cool-temperate forest communities and grassland. Furthermore, species richness decreased in subalpine *Abies* forests, riparian *Fraxinus* forests, and deciduous *Quercus* forests. If the damage continues for a long time after this stage, the entire forest structure changes (Gill 1992; Kirby and Thomas 2000); therefore, to prevent this, it is necessary to suppress the influence from deer (Akashi 2009).

In a case study on the use of fences to protect cool-temperate natural forests where species composition had already been altered by sika deer grazing, the original vegetation populations recovered. After deer grazing pressure was eliminated by protective fences, the tall perennial herbs and endangered species increased, and tree saplings and dwarf bamboo recovered (Tamura 2007, 2013). Deer protection fences have been used in many areas because they effectively facilitate the recovery of vegetation (Tamura 2008, 2013). However, it is important to quantify the impact of deer grazing on various types of forest vegetation to determine which of these are the most susceptible to the presence of deer. This facilitates determination of areas that need to be fenced.

Riparian forests established along rivers generally show high species diversity (Chap. 1). In particular, many herbaceous plants and ferns grow on the riparian forest floor and form forest floor vegetation unique to various micro-topographies formed by ground movements, such as debris flow and slope failure (Kawanishi et al. 2004, 2005, 2008). Forest floor vegetation within such species-rich forests must be taken into account, because it is susceptible to drastic changes due to the influence of deer.

8.2   Increase in Sika Deer Population and the Change in the Structure of an Ooyamazawa Riparian Forest

Surveys have been conducted in the Ooyamazawa riparian forest since 1983, collecting data on forest floor vegetation, although not every year. Population surveys of sika deer have been conducted while surveying serow populations in 1982 (Saitama Museum of Natural History 1983), 1987, 1992, 2000–2001, 2008–2009 (Gunma, Saitama, Tokyo, Yamanashi and Nagano Prefectural Boards

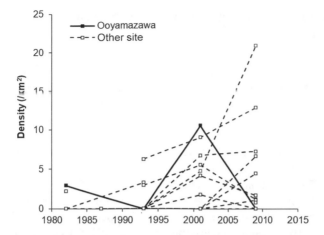

Fig. 8.1 Density of sika-deer population around area of Ooyamazawa. Solid square and open square indicate Ooyamazawa valley and the other investigation sites, respectively. The 1982 survey data were derived from Saitama Museum of Natural History (1983), and the data from 1987, 1992, 2000–2001, and 2008–2009 were derived from Gunma, Saitama, Tokyo, Yamanashi, and Nagano Prefectural Boards of Education (1988, 1994, 2002, 2010)

of Education 1988, 1994, 2002, 2010) in the vicinity of Ooyamazawa. These results show that the deer populations have been rapidly increasing since 2000 (Fig. 8.1). Although the presence of some deer was confirmed by Sakio in a 1982 serow survey, we seldom witnessed sika deer in the Ooyamazawa riparian forest between 1983 and 2000. Generally, the surveys did not report observation of feces or tree bark peeling during this period. Until 1999, the forest floor was covered with large ferns and herbs, and abundant *Fraxinus platypoda* and *Pterocarya rhoifolia* saplings also grew (Sakio 1997). However, this flora has disappeared from the forest floor since 2002. In addition, after 2000, we frequently spotted deer and heard cries at the time of investigation. In 2004, we found a deer carcass.

Sakio et al. (2013) reported changes in the forest floor vegetation of the Ooyamazawa riparian forest over 21 years from 1983 to 2004. The monitoring plot of this study is located at the bottom of a flat valley, which is approximately 60 m wide and buried in debris flow at an altitude of approximately 1300 m a.s.l. (Chap. 1). Plant species names and the degree of dominance in each stratum in the valley were recorded using the phytosociological vegetation survey method (Braun-Branquet 1964). The herbaceous layer measured 1 m or less, which is almost same as the height of large ferns (such as *Dryopteris crassirhizoma*, *Polystichum ovatopaleaceum*) and large herbs (such as *Cacalia yatabei*, *Spuriopimpinella nikoensis*). The four vegetation surveys conducted in 1983, 1998, 2001, and 2004 have been reported in Sakio et al. (2013). In 2016 and 2017, data were obtained in the same manner; these surveys were conducted from June to August in each survey year.

In this chapter, we would like to clarify the change in forest floor vegetation over 34 years by adding the new data from 2016 and 2017.

8.3   Changes in Coverage and Number of Species Over 34 Years

Forest floor vegetation within the Ooyamazawa riparian forest decreased precipitously in the 34 years from 1983 to 2017. The change was particularly significant from 1998 to 2004. In 1988, the plant cover ratio in the herbaceous layer reached about 90% because the forest floor was covered with large ferns and herbs as mentioned above (Fig. 8.2a). However, it decreased to approximately 50% in 2001, and it had plummeted to 3% by 2004 (Fig. 8.3). Through the above changes, most of the originally rich forest floor vegetation has currently disappeared (Fig. 8.2b). The reduction in herb cover followed the same timeline as the increase in deer population. The Ooyamazawa case exemplifies the characteristics of Japanese cold-temperate forest change brought on by deer that Ohashi (2017) pointed out. Even in 2016, which is 12 years after the alarming 2004 survey, the planting rate is less than 5%. Even in 2017, when the forest seemed to have recovered slightly, the cover rate is about 10%. We can recognize that the forest remains under strong grazing pressure from deer.

The number of species also decreased significantly during the same time span. We could observe 76 species in the herbaceous layer in 1983, but these decreased by

Fig. 8.2 (**a**) Physiognomy of forest floor vegetation in 1988 (photograph by Sakio, from Sakio et al. 2013). (**b**) Physiognomy of forest floor vegetation in 2003 (photograph by Sakio, from Sakio et al. 2013). Left and right picture shows the same site of Photo a, respectively

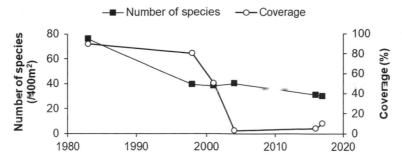

Fig. 8.3   Changes in coverage and number of species within the herb layer over 34 years

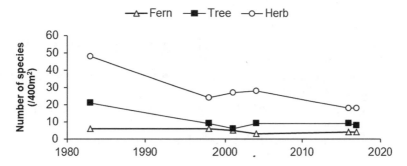

Fig. 8.4   Changes in number of species across each type of growth

half to 40 species in 2004, and further decreased to 30 species by 2017 (Fig. 8.3). The decreasing tendency showed a different pattern among the plant elements (Fig. 8.4). The plants confirmed during the survey period from 1983 to 2017 were 8 species of ferns, 24 woody species, and 57 herbaceous species, with a total number of 89 species. The number of fern species is relatively few and the change in the number of species cannot be clearly read, but the number of species lost has halved from six species to three species between 1983 and 2004, and most recently it was four species in 2017. Although the number of species lost can be as small as two or three species, it represents a significant loss framed within the proportion of total species. This decline in ferns had a significant influence in total vegetation cover in the herbaceous layer because several ferns often occupied large areas as the dominant species in the riparian forest floor (Kubo et al.2001; Kawanishi et al.2004, 2008). On the other hand, 21 woody species were confirmed in 1983, but decreased to nine species by 1998. Thereafter, only eight species were observed in 2017; there has been no major change since 2004. Although the cover ratio of saplings, except *Fraxinus platypoda*, was not large in the herbaceous layer, vanishing numbers of various woody saplings spells the loss of a seedling bank. This is seriously damaging to the regeneration of a forest. The herbaceous plants showed the largest reduction in both number of species number and proportion. There were 48 species of herbaceous plants in 1983, but this number decreased by half (24 species) in 1998 and then further decreased to 18 species by 2016.

8.4   Changes in Species Composition

The changing species patterns have been classified into six types. These types were made based on the fluctuation patterns seen from 1983 to 2004 reported by Sakio et al. (2013), and the variation in vegetation seen in 2016 and 2017 was added to complete the picture. The dominance of each species in each division is shown in Table 8.1.

"Changeless species": Species that appeared and did not change their dominance throughout the survey period.

"Species declined since 2001": Species that appeared throughout the study period, but dominance declined from 2001-onward.

"Disappeared species as of 2017": Species that were confirmed until 2004 or 2016, but were not present in 2017.

"Disappeared species as of 2004": Species that were confirmed until 1998 or 2001, but were not present in 2004.

"Species appeared after 2016": we could not confirm its presence before 2004, but could be observed after 2016.

"Other species": these species did not have a clear tendency of appearance over time.

There was a tendency for "Changeless species" to include the toxic plants (such as *Aconitum sanyoense*: Fig. 8.5, *Scopolia japonica*: Fig. 8.6) and the perennial herbs that wither above-ground early in the summer (such as *Adoxa moschatellina*: Fig. 8.7, *Cardamine leucantha*, *Trillium* spp.: Fig. 8.8, *Scopolia japonica*). In addition, perennial herbs distributed in unstable sites such as new collapsed areas and terraced scarp, such as *Elatostema umbellatum* var. *majus*: Fig. 8.9 and *Cacalia farfaraefolia* (see Chap. 6) were also included in this group.

The group of "Species declined since 2001" includes herbs (e.g., *Meehania urticifolia* and others) and ferns (e.g., *Cornopteris crenulatoserrulata*, *Polystichum tripteron*: Fig. 8.10), and seedlings of canopy tree species (*Fraxinus platypoda*, *Pterocarya rhoifolia*, *Abies homolepis*) that dominated in 1983. As seedlings of these tree species are regenerated by canopy trees almost every year, they might appear continuously.

Although the coverage of *Fraxinus platypoda* and *Pterocarya rhoifolia* in 1983 was large and they made dense sapling patches on the sand-gravel land along the ravine channel stream (Sakio 1997), their presence was greatly reduced by the influence of deer.

"Disappeared species as of 2017" includes species that had been eliminated at the time of survey in 2017 though included in the "declined species" category by Sakio et al. (2013). *Dryopteris crassirhizoma* (Fig. 8.11), *Asarum caulescens* (Fig. 8.12), *Mitella pauciflora*, etc. remained (despite their remarkable reduction in coverage)

Table 8.1 Changes in species composition over 34 years

	Lifeform	1983	1998	2001	2004	2016	2017
Changeless species							
Aconitum sanyoense	H	+		+2	+	1-2	+
Adoxa moschatellina	H	+	+	+2	+	+	+
Cucullu furfuraefolia	H	+	+	+	+		+
Cardamine leucantha	H	+	+	+	+	+	+
Chrysosplenium echinus	H	+		1-2	+	+	+
Elatostema umbellatum var. *majus*	H	+2	+2	+	+	+	+
Galium paradoxum	H	+	+	+2	+	+	+
Schizophragma hydrangeoides	T	+2	+2	+		+	+
Scopolia japonica	H	2-2	1-2	2-2	1-2	1-2	1-2
Thalictrum filamentosum var. *tenurum*	H	+	+		+	+	+
Trillium spp.	H	+	+	+	+		+
Declined species since 2001							
Abies homolepis	T	1-2				+	+
Cornopteris crenulatoserrulata	F	2-3	2-3	2-2	+	+	+
Fraxinus platypoda	T	3-3	+2	+	+	1-1	2-2
Hydrangea serrata	T	+	1-2	+	+	+	+
Laportea bulbifera	H	1-2	1-2	+	+		+
Meehania urticifolia	H	1-2	1-2	+2	+		+
Polystichum tripteron	F	1-2	+	+2	+	+	+
Pterocarya rhoifolia	T	1-2	+2		+	+	+
Disappeared species as at 2017							
Acer carpinifolium	T	+	+		+		
Acer nipponicum	T	+		+	+		
Asarum caulescens	H	2-2	2-2	1-2	+		
Cacalia delphiniifolia	H	1-2		+	+		
Caulophyllum robustum	H	+	+	+	+		
Chrysosplenium pilosum var. *sphaerospermum*	H	+		+2	+		
Deinanthe bifida	H	1-2	+2	+	+		
Dryopteris crassirhizoma	F	2-3	2-3	1-2	+		
Impatiens noli-tangere	H	+	+	+2	+		
Laportea macrostachya	H	1-2		+	+		
Mitella pauciflora	H	2-2	2-2	1-2	+		
Polystichum ovato-paleaceum	F	1-2	+	+		+	
Sambucus racemosa ssp. *sieboldiana*	T	+	+		+		
Smilacina japonica	H	1-2			+		
Spuriopimpinella nikoensis	H	1-2	+2	1-2	+	+	
Disappeared species as at 2004							
Acer mono	T	+	+	+			
Acer rufinerve	T	+	+				
Athyrium wardii	F	+2	+2				
Bistorta tenuicaulis	H	+	+				

(continued)

Table 8.1 (continued)

	Lifeform	1983	1998	2001	2004	2016	2017
Cacalia tebakoensis	H	+2	+2				
Cacalia yatabei	H	2-2	1-2	+2			
Cimicifuga simplex	H	+	+				
Clematis japonica	H	+	+	+			
Dryopteris polylepis	F	2-3	+	+			
Hydrangea petiolaris	T	+	+2				
Osmorhiza aristata	H	+	+				
Rodgersia podophylla	H	+	+				
Appeared species after 2016							
Astilbe thunbergii var. *thunbergii*	H					+	+
Chrysosplenium flagelliferum	H					+	+2
Chrysosplenium ramosum	H					+	+
Deparia pycnosora	F					+	+
Enemion raddeanum	H					+	+
Other species							
Acer amoenum	T					+	
Acer argutum	T	+				+	
Acer micranthum	T					+	
Acer shirasawanum	T	+					
Aconitum loczyanum	H	+			+		+
Actinidia arguta	T	+			+	+	+
Anemonopsis macrophylla	H	+					
Angelica polymorpha	H	+	+	+			+
Arisaema ovale var. *sadoense*	H					+	
Arisaema tosaense	H	+		+			
Bistorta suffulta	H	+		+			
Cardiocrinum cordatum	H	+		+			
Carpinus cordata	T	+					
Cercidiphyllum japonicum	T	+					+
Chelonopsis moschata	H	+					
Chrysosplenium album var. *stamineum*	H					+	
Chrysosplenium macrostemon var. *atrandrum*	H	+		+2			
Chrysosplenium ramosum	H	+			+		
Cimicifuga acerina	H	+					
Euonymus melananthus	T	+					
Euonymus oxyphyllus	T	+					
Fagus crenata	T						+
Fraxinus apertisqamifera	T	+					
Lamium album var. *barbatum*	H	+					
Lepisorus ussuriensis var. *distans*	F						+
Panax japonicus	H	+					

(continued)

Table 8.1 (continued)

	Lifeform	1983	1998	2001	2004	2016	2017
Paris verticillata	H	+					
Persicaria debilis	H	+			+		
Philadelphus satsumi	H	+		+			
Pilea hamaoi	H					+	
Pseudostellaria palibiniana	H	+			+		
Rabdosia umbrosa var. *leucantha*	H	1-2					
Rhamnus costata	T	+			+		
Smilax riparia var. *ussuriensis*	H	+					
Stellaria sessiliflora	H	+			+	+	
Tricyrtis latifolia	H					+	
Ulmus laciniata	T	+			+		
Veratrum album ssp. *oxysepalum*	H	+			+		

Lifeform abbreviations as follows: *H* herbs, *T* trees, *F* ferns. Numerical values in the table show Braun-Blanquet scale derived from phytosociological survey

Fig. 8.5 New leaves (2003/4/17 Ooyamazawa) and flowers (2002/8/25 Kamikochi, Nagano pref.) of *Aconitum sanyoense*

through 2004 and had disappeared by 2016. *Acer carpinifolium*, which was a major dominant shrub in the shrub layer of the riparian forest, had also vanished by 2017.

"Disappeared species as of 2004" are included in "disappeared species" in Sakio et al. (2013). We can recognize that these species disappeared at an early stage of deer grazing. Mainly small coverage species (such as *Cimicifuga simplex*, *Clematis japonica*, *Cacalia tebakoensis*) were included in this category, but high-coverage herbs from the 1983 survey such as *Cacalia yatabei* (Fig. 8.13) and *Dryopteris polylepis* were also included. In the case of tree species, *Acre mono* constitutes the tree layer of this riparian forest with a similar pattern.

The plants from the above-mentioned "disappeared species" are characterized as herbaceous and fern plants with tall stems (such as *Deinanthe bifida* (Fig. 8.14),

Fig. 8.6 *Scopolia japonica* (2002/4/18 Ooyamazawa)

Fig. 8.7 *Adoxa moschatellina* (2010/5/9 Ooyamazawa)

Cacalia delphiniifolia, *Impatiens noli-tangere* (Fig. 8.15), and *Laportea macrostachya* (Fig. 8.16)) or large leaf (*Cacalia yatabei*, *Caulophyllum robustum*, *Polystichum ovato-paleaceum*, and *Dryopteris polylepis*).

On the other hand, the group of "Species appeared after 2016" contained five confirmed species: *Astilbe thunbergii* var. *thunbergii*, *Chrysosplenium flagelliferum*, *Chrysosplenium ramosum*, *Deparia pycnosora*, and *Enemion raddeanum*. However, we cannot conclude whether these could be attributed to the decline in forest floor

Fig. 8.8 *Trillium tschonoskii* (2003/5/16 Ooyamazawa)

Fig. 8.9 *Elatostema umbellatum* var. *majus* (2016/7/24 Ooyamazawa)

vegetation or not, because these species have a coverage rate of less than 1%, even in 2017. The possibility of coincidence cannot be denied.

As mentioned above, the increase in the deer population had such an influence as to drastically alter the forest floor vegetation of the Ooyamazawa riparian forest. The

Fig. 8.10 *Polystichum tripteron* (2016/11/26 Gero, Gifu Pref.)

Fig. 8.11 *Dryopteris crassirhizoma* (2014/9/13 Kamikochi, Nagano Pref.)

Fig. 8.12 *Asarum caulescens* (2011/9/28 Nikko, Tochigi Pref.)

Fig. 8.13 Diminishment of *Cacalia yatabei* leaves due to constant grazing (2016/7/24 Ooyamazawa)

Fig. 8.14 *Deinanthe bifida* (2016/7/23 Ooyamazawa)

plant cover rate changed from 90% to 10% or less, and the number of emerging species also declined precipitously from 76 to 30 species. It was also revealed that such a drastic change in the forest floor vegetation occurred in a short period of several years.

8.5   Ecological Characteristics of Forest Floor Vegetation After Deer Impact

Ecological characteristics were observed in the forest floor vegetation under the continued pressure of deer grazing. Considering the remaining species as of 2017, the plants had one of the following characteristics.

Fig. 8.15 *Impatiens noli-tangere* (2016/7/24 Ooyamazawa)

Fig. 8.16 *Laportea macrostachya* (2016/7/24 Ooyamazawa)

Fig. 8.17 Flowers of *Veratrum album* ssp. *oxysepalum* (2016/7/23 Ooyamazawa) and large population in area under deer grazing pressure (2003/5/18 Tanzawa, Kanagawa pref.)

8.5.1   Toxic Plants

For example, *Aconitum sanyoense*, *Scopolia japonica*, *Veratrum album* ssp. *oxysepalum* (Fig. 8.17). This trait is recognized as unpalatability. An increase in unpalatable plants was the most frequently reported in the world and a significant relationship has been documented between grazing and palatability (Diaz et al. 2007).

8.5.2   Small Herb Species that do not Increase in Plant Height

For example, small size herbs such as *Galium paradoxum*, *Adoxa moschatellina*, and *Pseudostellaria palibiniana* and creeping plants such as those of the *Chrysosplenium* species fit this description (Figs. 8.18, 8.19, and 8.20). Plant size and growth form are important and effective for estimating deer feeding probability (Diaz et al.2007; Takatsuki 2015). Herbaceous plants with heights of 30 cm or more which are upright, branched, or on vines are susceptible to feeding and showed larger impacts (Takatsuki 2015). On the other hand, small species such as rosettes and creeping plants have an easier time avoiding deer. Ohashi et al. (2007) showed that the small herbs increase or do not change under deer grazing. Kirby and Thomas (2000) show that the species that increased had the tendency to be more ruderal in Grime's strategy; they were smaller and less likely to have leafy stems although the differences were not significant.

Fig. 8.18 *Chrysosplenium echinus* (2002/4/18 Ooyamazawa)

Fig. 8.19 *Chrysosplenium macrostemon* (2002/4/18 Ooyamazawa)

8.5.3   Perennial Herbs in Which the Aerial Portion Withers Until the Summer and Enters Yearly Dormancy at an Early Date

For example, *Adoxa moschatellina*, *Trillium* spp., and *Cardamine leucantha*. Spring ephemerals plants are representative of this trait. The data of our survey did not include spring ephemerals because the seasons of growth differed. In observations by Sakio in the early spring after snow melting, it was confirmed that two spring ephemeral species, *Corydalis lineariloba* (Fig. 8.21) and *Allium monanthum* (Fig. 8.22) have maintained their populations without serious deer damage.

Fig. 8.20   *Chrysosplenium pilosum* var. *sphaerospermum* (2002/4/18 Ooyamazawa)

Fig. 8.21   *Corydalis lineariloba* (2010/5/8 Ooyamazawa)

In this way, plants containing toxic components, small plants, and plants with short growing periods that are difficult to eat whole have remained prolific in the current forest floor vegetation.

The annual grasses, such as *Impatiens noli-tangere* and *Persicaria debilis*, have survived through 2004; they are generally thought to be more resistant to grazing pressures than perennials (Diaz et al.2007). Therefore, these annual species were recognized as some of the remaining species by Sakio (2013). However, these species have not been confirmed in the most recent survey. It is expected that the life history characteristics of these species are not sufficient to maintain the population under strong grazing pressures. Both plants might be frequently eaten because

Fig. 8.22 *Allium monanthum* (2010/5/8 Ooyamazawa)

they can grow to or over 30 cm in stem height, and then their flowers and fruits are also lost. As a result, seed dispersal had might be restricted. In addition, seed banks were not rich (Kubo et al.2008, seed dormancy is unknown). These factors may be related to the decline in both species.

There are many reports that forest floor vegetation declines due to increasing deer density in many forests, and vegetation is simplified by the proliferation of unpalatable plants in Japan (Tamura 2007; Hattori et al.2010; Ishida et al.2012; Otsu et al.2011; Shimoyama 2012) and the rest of the world (Diaz et al.2007). Furthermore, when there is a decrease in edible plants, unpalatable plants are also eaten (Ishikawa 2010). In the Ooyamazawa riparian forest, the cover ratio of plants declined rapidly in the 6 years from 1998 to 2004, and then the number of species further decreased after 2004 (Fig. 8.2). Most palatable plants were eaten by 2004, and the number of species seemed to increase temporarily afterwards, but it seems that the number of unfavorable species also decreased until 2016. It has been recorded that poisonous, diminutive, and spring ephemeral plants are avoided (Diaz et al.2007) as mentioned above. The future change in the plant cover will depend on whether these species will be able to survive or recover in the future. On the other hand, forest regeneration will be very difficult because tree seedlings will be eliminated from the forest floor if damage by deer grazing continues over a long period of time (Takatsuki and Gorai 1994). At our study site, the seedlings of the dominant canopy tree species have continuously appeared from the seeds supplied by the canopy. However, the patch of saplings from the dominant species is disappearing, and seedlings from some tree species, especially the maple family, have not appeared. For this reason, some tree species may disappear from the forest floor entirely.

8.6   Conservation of Species Diversity Within the Forest Floor Vegetation

Ideally, rare plants and communities should be conserved by managing the populations of deer and/or installing deer fences before the number of deer increases (Ōsawa et al.2015). Protection of plant communities through the installation of deer fences is an effective method under the conditions where high deer densities and strong grazing pressures are present (Tamura 2008). However, it is important to carefully select installation points within the protected areas because the cover area of the deer fence is relatively small.

Species that disappeared or decreased since 2001 in the surveyed area are ferns with large leaves such as *Cornopteris crenulatoserrulata*, *Dryopteris crassirhizoma*, and *Polystichum ovato-paleaceum*, high stem grasses such as *Caulophyllum robustum*, *Spuriopimpinella nikoensis*, and shrubs like *Hydrangea serrata* and *Sambucus racemosa* ssp. *Sieboldiana*, etc. These species tend to be distributed in relatively stable habitats like the debris flow terraces in Ooyamazawa (Kawanishi et al.2004, Chap. 6). Such locations can be also an update site for species such as *Fraxinus platypoda* (Sakio 1997), so it is desirable to preferentially protect this community in this habitat.

However, as mentioned above, the forest floor vegetation of the Ooyamazawa riparian forest has already undergone significant changes, and the species diversity has been greatly reduced. *Sasa borealis*, which covered the forest floor at the upper part of the slope, has started to die in large areas (Fig. 8.23). Therefore, protecting the remaining individuals is very important. In the investigation area, large herbs and shrubs, such as *Angelica polymorpha* and *Hydrangea serrata* remained only on large riverbed rocks (Fig. 8.24). This is a location unaffected by grazing because deer cannot reach the rocky riverbed. In contrast, outside the survey area, the rocky cliffs along the valley slope are also noteworthy (Fig. 8.25). Although the species that can grow in such places seem to be limited, these habitats seem to be functioning as

Fig. 8.23 Dead patch of *Sasa borealis* due to the deer grazing (2017/7/22 Ooyamazawa)

Fig. 8.24 Fragmentary community of large herbs remained on the large rocks of the riverbed (2016/7/24 Ooyamazawa)

Fig. 8.25 Community of large herbs established on the rocky cliff along the stream. (2017/7/22 Ooyamazawa)

reprieves from deer grazing for some high stem herbs such as *Deinanthe bifida* and *Elatostema umbellatum* var. *majus*. When installing a deer fence, it would be desirable to position it in a manner that facilitated proliferation of seeds from these plants.

In Ooyamazawa, a deer fence was installed at the valley slope in 2017 (Fig. 8.26). We could find many herbs within the deer fence. As of 2017, they are found in the steep sloping rock walls only. However, these herbs had also grown in the riparian forest originally; therefore, we expect a recovery in the forest floor vegetation within

Fig. 8.26 Deer fences installed in Ooyamazawa since 2017 (photograph-A by Kubo in spring 2018, photograph-B in 2017/7/22 Ooyamazawa)

the fence. In my future studies, I would like to focus on changes in vegetation within and outside the deer fence.

References

Akashi N (2009) Simulation of the effects of deer browsing on forest dynamics. Ecol Res 24:247–255

Braun-Branquet J (1964) Pflanzensoziologie, grundzuge der vegetationskunde, 3rd edn. Springer, Wien, p 865

Diaz S, Lavorel S, McIntyre AS, Falczuk V, Casanoves F, Milchunas DG, Skarpe C, Rusch G, Sternberg M, Noy-Meir I, Landsberg J, Zhang W, Clark H, Campbell B (2007) Plant trait responses to gradzing – a global synthesis. Glob Chang Biol 13:313–341

Gill RMA (1992) A review of damage by mammals in north temperate forests: 3. Impact on trees and forests. Forestry 65:363–388

Gunma, Saitama, Tokyo, Yamanashi and Nagano Prefectural Boards of Education (1988) Bulletin report of Japanese Serow (*Capricornis crispus*) protection area in Kanto Mountains. Gunma, Saitama, Tokyo, Yamanashi and Nagano Prefectural Boards of Education, Tokyo, p 102

Gunma, Saitama, Tokyo, Yamanashi and Nagano Prefectural Boards of Education (1994) Bulletin report of Japanese Serow (*Capricornis crispus*) protection area in Kanto Mountains. Gunma, Saitama, Tokyo, Yamanashi and Nagano Prefectural Boards of Education, Tokyo, p 113

Gunma, Saitama, Tokyo, Yamanashi and Nagano Prefectural Boards of Education (2002) Bulletin report of Japanese Serow (*Capricornis crispus*) protection area in Kanto Mountains. Gunma, Saitama, Tokyo, Yamanashi and Nagano Prefectural Boards of Education, Tokyo, p 129

Gunma, Saitama, Tokyo, Yamanashi and Nagano Prefectural Boards of Education (2010) Bulletin report of Japanese Serow (*Capricornis crispus*) protection area in Kanto Mountains. Gunma, Saitama, Tokyo, Yamanashi and Nagano Prefectural Boards of Education, Tokyo, p 152

Hattori T, Tochimoto D, Minamiyama N, Hashimoto Y, Fujiki D, Ishida H (2010) Influence of feeding pressure by Sika deer (*Cervus nippon*) on the primeval lucidophyllous forest in Kawanaka, Aya, Miyazaki Prefecture. Veg Sci 27:35–42

Ishida H, Hattori T, Kuroda A, Hashimoto Y, Iwakiri K (2012) Assessment of the effects of Sika deer (*Cervus nippon yakushimae*) on secondary lucidophyllous forests and an evaluation of forest naturalness in lowland areas of Yakushima Island, Japan. Veg Sci 29(1):49–72

Ishikawa Y (2010) Grazing aspect of Yeso Shika-deer at deciduous broad-leaved forest in Hokkaido. Veg Sci News 14(3):9–12

Kawanishi M, Sakio H, Ohno K (2004) Forest floor vegetation of *Fraxinus platypoda-Pterocarya rhoifolia* forest along Ooyamazawa valley in Chichibu, Kanto District, Japan, with a special reference to ground disturbance. Veg Sci 21:15–26

Kawanishi M, Ishikawa S, Miyake N, Ohno K (2005) Forest floor vegetation of Pterocarya rhoifolia forests in Shikoku, with special reference to micro-landform. Veg Sci 22:87–102

Kawanishi M, Sakio H, Ohno K (2008) Diversity of forest floor vegetation with landform type. In: Sakio H, Tamura T (eds) Ecology of riparian forests in Japan: disturbance, life history, and regeneration. Springer, Tokyo, pp 267–278

Kirby KJ, Thomas RC (2000) Changes in the ground flora in Wytham Woods, southern England from 1974 to 1991 – implications for nature conservation. J Veg Sci 11:871–880

Kubo M, Shimano K, Ohno K, Sakio H (2001) Relationship between habitats of dominant trees and vegetation units in Chichibu Ohyamasawa riparian forest. Veg Sci 18:75–85

Kubo M, Kawanishi M, Shimano K, Sakio H, Ohno K (2008) The species composition of soil seed banks in the Ooyamazawa riparian forest, in the Chichibu Mountains, Central Japan. J Jpn Forest Soc 90:121–124

Ohashi H (2017) The effects of Sika deer on natural vegetation. In: Kaji K, Iijima H (eds) Sika deer in Japan. The science and management of overabundant populations. University of Tokyo Press, Tokyo, pp 29–45

Ohashi H, Hoshino Y, Oono K (2007) Long-term changes in the species composition of plant communities caused by the population growth of Sika deer (Cervus nippon) in Okutama, Tokyo. Veg Sci 24:123–151

Ohashi H, Yoshikawa M, Oono K, Tanaka N, Hatase Y, Murakami Y (2014) The impact of Sika deer on vegetation in Japan: setting management priorities on a national scale. Environ Manag 54:631–640

Osawa T, Igehara G, Ito C, Michimata S, Sugiyama D (2015) A method for the early detection of vegetation degradation caused by Sika deer. Jpn J Conserv Ecol 20:167–179

Otsu C, Hoshino Y, Matsuzaki A (2011) Impacts of the Sika deer on montane and subalpine grasslands in Chichibu-Tama-Kai National Park and surrounds, central Japan. Veg Sci 28:1–17

Planning Committee-The Society of Vegetation Science (2011) Impacts of Sika deer (*Cervus nippon*) on Japanese vegetation surveyed by questionnaire in 2009-2010. Veg Sci News 15:9–96

Saitama Museum of Natural History (1983) Bulletin report of special natural monument: density of Japanese Serow. Saitama Prefectural Boards of Education, Saitama, p 27

Sakio H (1997) Effects of natural disturbance on the regeneration of riparian forests in a Chichibu mountains, central Japan. Plant Ecol 132(2):181–195

Sakio H, Kubo M, Kawanishi M, Higa M (2013) Effects of deer feeding on forest floor vegetation in the Chichibu mountains, Japan. J Jpn Soc Reveget Tech 39:226–231

Shimoyama Y (2012) Species composition of secondary forests browsed by Sika deer in Oshika Peninsula of Miyagi Prefecture. Veg Sci 29:111–117

Takatsuki S (2009) Effects of Sika deer on vegetation in Japan: a review. Biol Conserv 142:1922–1929

Takatsuki S (2015) The effects of Sika deer on plants. In: Maesako Y, Takatsuki S (eds) Deer's threat and the future of the forest. Effectiveness and limit of vegetation conservation by deer fence. Bun-ichi Sogo Shuppan, Tokyo, pp 31–41

Takatsuki S, Gorai T (1994) Effects of Sika deer on the regeneration of a Fagus crenata forest on Kinkazan Island, northern Japan. Ecol Res 9:115–120

Tamura A (2007) Changes in understory vegetation over 10 years after the exclusion of Sika deer browsing pressure in cool temperate deciduous forests, central Japan. Jpn J Fer Environ 49 (2):103–110

Tamura A (2008) Effects of fencing for 7 years on the regeneration of tree species after dwarf bamboo on the forest floor has been diminished by Sika deer browsing. J Jpn Forest Soc 90:158–165

Tamura A (2013) Effects of time lag in establishment of deer-proof fences on recovery of floor vegetation and regeneration of tall trees in a beech forest diminished by Sika deer browsing in the Tanzawa mountains of central Japan. J Jpn Forest Soc 95:8–14

Chapter 9
Temporal Changes in Browsing Damage by Sika Deer in a Natural Riparian Forest in Central Japan

Motoki Higa, Motohiro Kawanishi, Masako Kubo, and Hitoshi Sakio

Abstract Over the last few decades, population increases of sika deer (*Cervus nippon*) have become a major issue in various forest ecosystems across temperate regions; however, the influences of deer browsing on riparian forests are less known. In this chapter, we illustrate the herbivore–forest vegetation relationships over a long term from the past when deer was absent to the current when deer became overabundant in an old-growth riparian forests of Ooyamazawa, the Chichibu Mountains of central Japan. We revealed that (1) the browsing activity of deer has a negative influence on riparian forests, (2) the damage of these species is mainly induced by easiness to browsing by deer resulting from small tree size structure, and (3) the resistance to the deer browsing differs among tree species. Thus, small mature trees (i.e., shrub species) with low browsing resistance should be primarily protected for effective management of riparian forests.

Keywords *Cervus nippon* · Debarking · Forest management · Mortality risk · Species preferences · Tree size

M. Higa (✉)
Faculty of Science and Technology, Kochi University, Kochi, Japan
e-mail: mhiga@kochi-u.ac.jp

M. Kawanishi
Faculty of Education, Kagoshima University, Kagoshima, Japan
e-mail: kawanishi@edu.kagoshima-u.ac.jp

M. Kubo
Faculty of Life and Environmental Science, Shimane University, Shimane, Japan
e-mail: kubom@life.shimane-u.ac.jp

H. Sakio
Sado Island Center for Ecological Sustainability, Niigata University, Niigata, Japan
e-mail: sakio@agr.niigata-u.ac.jp; sakiohit@gmail.com

© The Author(s) 2020
H. Sakio (ed.), *Long-Term Ecosystem Changes in Riparian Forests*, Ecological Research Monographs, https://doi.org/10.1007/978-981-15-3009-8_9

9.1   Introduction

Browsing by large herbivores can drastically modify the structure, environment, and tree-species composition of forests (Gill 1992a, b). Over the last few decades, sika deer (*Cervus nippon*) populations have increased due to the absence of predators and a gradual decline in hunting, and have become a major issue for forest ecosystems in several regions of Japan (Uno et al. 2007; Takatsuki 2009; Kaji et al. 2010). In places where deer density is too high, plant-species diversity has decreased due to a drastic decline in forest-floor vegetation (Suda et al. 2001; Suzuki et al. 2008) and the increased mortality of palatable plant species (Akashi and Nakashizuka 1999; Yokoyama et al. 2001) caused by heavy browsing and debarking activities of deer. Thus, the adverse effects of these activities on the regeneration of tree species in natural forests have become a major concern (Takatsuki and Gorai 1994; Akashi and Nakashizuka 1999; Suda et al. 2001; Yokoyama et al. 2001). Previous studies have shown that increasing deer populations can also have other impacts on forest ecosystems and resources. For example, they may reduce the quality of timber resources (Kaji et al. 2000; Oi and Suzuki 2001; Ueda et al. 2002; Akashi and Terazawa 2005; Suzuki et al. 2008), decrease the diversity of insect communities (Kanda et al. 2005; Ueda et al. 2009), alter the structure and function of soil microbial food webs (Niwa et al. 2011), and increase the risks of soil erosion and landslides in mountainous areas (Furusawa et al. 2003).

An understanding of herbivore–vegetation relationships is necessary for the assessment of potential risks and design of management plans. In Japan, deer consume forest-floor vegetation and tree leaves from spring to autumn and utilize tree bark and dead fallen leaves when facing food shortages, which usually occur in winter (Takahashi and Kaji 2001; Takatsuki 2009). Browsing damage from deer occurs intensively on certain plant species among a wide array of co-occurring species (Gill 1992a; Augustine and McNaughton 1998; Côté et al. 2004; Boulangera et al. 2009). Tree-species composition and food-resource availability in forests affect the dietary selections of deer (Gill 1992a; Moser et al. 2006; Jayasekara and Takatsuki 2000; Takahashi and Kaji 2001). Many researchers have proposed factors to explain differences in the plant-species preferences of deer, including the morphological and physiological traits of plants (Gill 1992a; Ando et al. 2003; Jiang et al. 2005; Moser et al. 2006; Sauvé and Côté 2006). For example, smaller trees are well known to be preferentially browsed by deer (Gill 1992a; Boulangera et al. 2009; Didion et al. 2009) because they have more leaves and branches at heights easily accessible to deer, as well as thin and soft bark. The trunk size of these trees is also suitable for antler fraying. Large-herbivore densities also affect the occurrence and intensity of browsing damage on plants (Gill 1992a). However, the temporal changes in dietary selection and browsing damage on plants associated with deer density changes have been poorly documented. In particular, information about the initial changes that occur when an area transitions from the absence to overabundance of deer is lacking.

Riparian forests are important and unique ecosystems that provide specialized habitats and resources for wildlife, support and maintain faunal and floral diversity, and prevent soil erosion in precipitous mountain terrain (Malanson 1993; Richards et al. 2002; Ward et al. 2002). The majority of riparian forests in Japan disappeared during the twentieth century due to increased industrial or agricultural land use and the construction of embankments for river management (Sakio 2008); several of the remaining forests are being subjected to heavy browsing pressure by deer. In an old-growth riparian forest dominated by *Ulmus davidiana* Planch. var. *japonica* (Rehder) Nakai in the district of Nikko in central Japan, heavy browsing and debarking by deer have resulted in the decline of forest-floor vegetation and dieback of mature trees (Nomiya et al. 2003). The deer population is expected to continue to increase and expand throughout the less-snowy regions of Japan. Hence, riparian forests are expected to be subject to increasingly serious damage due to deer browsing activity.

In this chapter, we illustrate long-term herbivore–forest vegetation relationships in an old-growth riparian forest of Ooyamazawa, in the Chichibu Mountains of central Japan, from the past, when deer were absent, to the present, when deer are overabundant. Using long-term tree-census data, we elucidated the temporal changes in the dietary selections of deer and species-specific differences in browsing damage. Finally, we discuss effective management plans for riparian forests.

9.2   Ooyamazawa Riparian Forest

In Ooyamazawa (35°57′30″N, 138°46′32″E, 1200–1620 m a.s.l.), an old-growth natural forest lies along a mountain stream of the Arakawa River in part of Chichibu-Tama-Kai National Park, Saitama Prefecture, central Japan (Fig. 9.1). The annual mean temperature, annual precipitation, and maximum snow depth in the nearest settlement (Nakatsugawa; 700 m a.s.l. and 4.6 km from the study site) are 10.7 °C, 1100 mm, and 30 cm, respectively (Sakio 1997). The estimated annual mean temperature at the study site (1450 m a.s.l.) is 6.5 °C, based on a temperature lapse rate of 0.6 °C per 100-m increase in elevation (Sakio 1997). The site lies in the upper part of a cool-temperate, deciduous broad-leaf forest zone (Sakio 1997). The dominant species of the forest canopy are *Flaxinus platypoda* Oliv., *Pterocarya rhoifolia* Siebold et Zucc., and *Cercidiphyllum japonicum* Siebold et Zucc. ex Hoffm. et Schult., which are more than 30 m in height (Table 9.1). The subcanopy is dominated by *Acer shirasawanum* Koidz. and *Acer pictum* Thunb. The understory is composed primarily of *Acer carpinifolium* Siebold et Zucc. and *Acer argutum* Maxim (Table 9.1). Before deer became abundant, the forest floor had dense vegetation cover with greater plant-species richness, including *Parasenecio tebakoensis* (Makino) H.Koyama, *Parasenecio delphiniifolius* (Siebold et Zucc.) H.Koyama, *Chrysosplenium macrostemon* Maxim. var. *macrostemon*, *Laportea bulbifera* (Siebold et Zucc.) Wedd., *Impatiens noli-tangere* L., *Dryopteris*

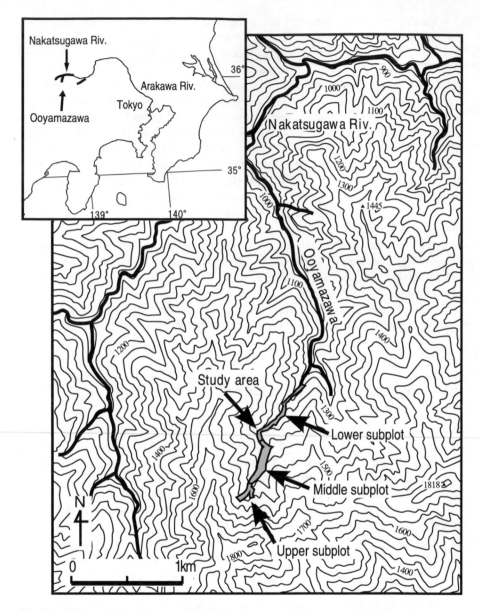

Fig. 9.1 Index and location maps of the study area

crassirhizoma Nakai, *Dryopteris polylepis* (Franch. et Sav.) C.Chr., and *Polystichum tripteron* (Kunze) C.Presl (Sakio et al. 2013).

In the Ooyamazawa riparian forest, a permanent study plot (4.71 ha; Fig. 9.1) consisting of three subplots (upper, middle, and lower) was established, and tree censuses were conducted at each subplot at 5-year intervals from 1991 to 2008. The upper and middle subplots (3.2 ha) were established in 1991, and the lower subplot

Table 9.1 Tree-species composition, life form, and size structure of the Ooyamazawa riparian forest (4.71 ha^2)

Species	LF	N. ind	N. tree	BA	RBA	Max. DBH
Fraxinus platypoda Oliv.	D, B, T	460	463	88.61	55.2	140.5
Cercidiphyllum japonicum Siebold et Zucc. ex Hoffm. et Schult.	D, B, T	39	39	23.02	15.6	153.4
Pterocarya rhoifolia Siebold et Zucc.	D, B, T	118	118	14.07	8.8	77.7
Acer shirasawanum Koidz.	D, B, T	428	428	7.33	4.6	62.8
Acer pictum Thunb.	D, B, T	271	272	6.16	3.8	92.0
Ulmus laciniata (Trautv.) Mayr ex Schwapp.	D, B, T	95	95	6.32	3.9	89.4
Acer carpinifolium Siebold et Zucc.	D, B, S	484	508	2.82	1.8	47.0
Tilia japonica (Miq.) Simonk.	D, B, T	9	9	1.92	1.2	93.9
Betula maximowicziana Regel	D, B, T	4	4	1.42	0.9	73.0
Abies homolepis Siebold et Zucc.	E, C, T	12	12	1.70	1.1	93.0
Carpinus cordata Blume	D, B, T	52	52	0.76	0.5	21.6
Kalopanax septemlobus (Thunb.) Koidz.	D, B, T	3	3	0.84	0.5	78.0
Fagus crenata Blume	D, B, T	6	6	0.52	0.3	62.6
Betula grossa Siebold et Zucc.	D, B, T	4	4	0.52	0.3	50.8
Padus buergeriana (Miq.) T.T.Yü et T.C.Ku	D, B, T	1	1	0.31	0.2	62.7
Acer argutum Maxim.	D, B, S	138	141	0.37	0.2	13.2
Acer palmatum Thunb.	D, B, S	6	6	0.11	0.1	23.9
Acer rufinerve Siebold et Zucc.	D, B, S	11	11	0.18	0.1	33.6
Pterostyrax hispida Siebold et Zucc.	D, B, S	89	89	0.38	0.2	16.5
Phellodendron amurense Rupr.	D, B, T	1	1	0.10	0.1	35.5
Actinidia arguta (Siebold et Zucc.) Planch. ex Miq.	D, B, L	13	13	0.10	0.1	16.2
Acer maximowiczianum Miq.	D, B, T	2	2	0.14	0.1	39.0
Fraxinus lanuginosa Koidz. f. *serrata* (nakai) Murata	D, B, T	2	2	0.03	<0.1	19.5
Aria alnifolia (Siebold et Zucc.) Decne.	D, B, T	4	4	0.07	<0.1	25.5
Schizophragma hydrangeoides Siebold et Zucc.	D, B, L	8	8	0.03	<0.1	10.1
Acer amoenum Carriére var. *amoenum*	D, B, S	3	3	0.03	<0.1	15.5
Carpinus japonica Blume	D, B, T	1	1	0.00	<0.1	7.0
Tsuga sieboldii Carriére	E, S, T	2	2	0.07	<0.1	24.5
Hydrangea petiolaris Siebold et Zucc.	D, B, L	1	1	0.01	<0.1	8.0
Acer nipponicum H.Hara	D, B, T	16	16	0.15	0.1	28.7
Trochodendron aralioides Siebold et Zucc. f. *longifolium* (Maxim.) Ohwi	E, B, T	2	2	0.02	<0.1	13.2
Acer japonicum Thunb.	D, B, T	3	3	0.00	<0.1	3.9
Acer tenuifolium (Koidz.) Koidz.	D, B, T	7	7	0.04	<0.1	15.0
Euptelea polyandra Siebold et Zucc.	D, B, S	10	10	0.08	<0.1	16.0
Euonymus sieboldianus Blume	D, B, T	7	7	0.05	<0.1	20.0

(continued)

Table 9.1 (continued)

Species	LF	N. ind	N. tree	BA	RBA	Max. DBH
Swida controversa (Hemsl. ex Prain) Soják	D, B, T	2	2	0.08	<0.1	23.9
Acer cissifolium (Siebold et Zucc.) K.Koch	D, B, T	1	2	0.03	<0.1	18.2
Acer tschonoskii Maxim.	D, B, T	1	1	0.00	<0.1	4.1
Fraxinus apertisquamifera H.Hara	D, B, T	9	9	0.05	<0.1	15.3
Vitis coignetiae Pulliat ex Planch.	D, B, L	4	4	0.02	<0.1	10.0
Clethra barbinervis Siebold et Zucc.	D, B, T	3	3	0.01	<0.1	6.4
Viburnum furcatum Blume ex Maxim.	D, B, S	7	7	0.01	<0.1	6.8
Stewartia pseudocamellia Maxim.	E, B, T	2	2	0.04	<0.1	17.6
Padus grayana (Maxim.) C.K.Schneid.	D, B, T	1	1	0.10	0.1	35.0
Acer distylum Siebold et Zucc.	D, B, S	1	1	0.02	<0.1	14.0
Celtis jessoensis Koidz.	D, B, S	1	1	0.00	<0.1	4.1

LF life form, *D* deciduous, *E* evergreen, *B* broad-leaf, *C* needle-leaf, *T* tall tree, *S* small tree or shrub, *L* liana. *N. ind* number of individuals, *N. tree* number of standing trees, *BA* basal area (m^2), *RBA* relative basal area (%)

(1.5 ha) was established in 1998. All living trees with diameters at breast height (DBHs) > 4 cm were numbered and identified to the species level. Tree DBHs, alive/dead status, and extent of browsing damage by deer were recorded. An initial survey was conducted at the upper subplot in 1991 and at the middle subplot in 1992; 1982 standing trees of 42 species were observed and measured, and the total basal area was 36.3 m^2 ha^{-1}. A second survey was conducted in the upper, middle, and lower subplots from 1996 to 1998. The third and fourth surveys were conducted from 2001 to 2003 and 2006 to 2008, respectively. The second, third, and fourth surveys included 2396 standing trees of 46 species, 2335 standing trees of 45 species, and 2050 standing trees of 41 species, and the total basal areas were 34.2, 35.2, and 35.5 m^2 ha^{-1}, respectively. The entire permanent plot was designated as an associate site of the Japan Long-Term Ecological Research Network in 2006, and the middle subplot has been a core site of the nationwide Japanese Monitoring Sites 1000 program since 2008.

9.3 Increase in Deer Density Around Ooyamazawa

Deer density has been monitored using the block-count method for more than 30 years as part of density monitoring for a protected species, Japanese serow (*Capricornis crispus*) (Saitama Museum of Natural History 1983; Gunma, Saitama, Tokyo, Yamanashi and Nagano Prefectural Boards of Education 1988, 2010). During the 1980s and 1990s, the deer density around the study site was 0–6.3 heads/km². This density began to increasing in the late 1990s, and had reached 20.9 heads/km² in the 2000s (Saitama Museum of Natural History 1983; Gunma,

Saitama, Tokyo, Yamanashi and Nagano Prefectural Boards of Education 1988, 2010). The deer censuses were performed using the block-count method on only single days every few years. The visibility of the census site is well known to affect the accuracy of density estimation when using this method (Maruyama and Furubayashi 1983). In Ooyamazawa and its surroundings, the slope is very steep and the geomorphology is complex. Therefore, the actual densities may be greater than estimated. The deer are permanent, year-round residents of the forest.

9.4 Temporal Changes in Browsing Damage

Grazing and browsing of overabundant large herbivores drastically change the structure and species composition of forest vegetation (Ripple and Beschta 2008; Didion et al. 2009; Salk et al. 2011). By the 2010s, most herbaceous plants and the seedlings and saplings of all tree species had disappeared due to heavy deer foraging pressure (Sakio et al. 2013). On the forest floor, only unpalatable and poisonous plants, such as *Veratrum album* subsp. *oxysepalum* (Turcz.) Hultén, *Aconitum tonense* Nakai ex H.Hara, and *Scopolia japonica* Maxim., remain in low abundance (Sakio et al. 2013).

Many studies have shown that heavy browsing and debarking activities of deer result in failed regeneration of tree species and reduced plant-species diversity (Takatsuki and Gorai 1994; Akashi and Nakashizuka 1999; Suda et al. 2001; Yokoyama et al. 2001; Suzuki et al. 2008). Similar phenomena were observed in the Ooyamazawa riparian forest. In the first survey, no trace of browsing damage was found on any standing tree in the upper or middle subplot (Table 9.2). In the second survey, evidence of debarking was found on three trees of *U. laciniata* and one tree of *A. shirasawanum* (Table 9.3). Thereafter, evidence of deer browsing increased rapidly; it was observed on 170 trees of eight species in the third survey. Among the damaged species, *A. carpinifolium* and *A. argutum* showed considerably increased browsing damage, which affected approximately 20 and 25%, respectively, of the total number of trees of these species. The three dominant canopy species, *F. platypoda*, *C. japonicum*, and *P. rhoifolia*, exhibited no evidence of browsing damage. By the fourth survey, browsing damage had increased to 536 trees

Table 9.2 Temporal changes in browsing damage (ND: number of damaged trees) of sika deer on tree species in the Ooyamazawa riparian forest during 1991–2008

Subplot	First survey (1991–1992)		Second survey (1996–1998)		Third survey (2001–2003)		Fourth survey (2006–2008)	
	ND	TN	ND	TN	ND	TN	ND	TN
Upper	0	622	4 (0.7)	576	57 (11.5)	494	101 (25.3)	399
Middle	0	1361	0	1286	20 (1.7)	1183	267 (25.6)	1042
Lower	–	–	0	550	93 (13.8)	674	168 (27.5)	611

The percentage of damaged trees is shown in parentheses. *TN* total number of trees

Table 9.3 Temporal changes in browsing damage (number of damaged trees in 4.71 ha), mortality rate, and percentage of browsed trees among all dieback trees among the main 10 tree species in the Ooyamazawa riparian forest from the second through fourth surveys

	Second survey (1996–1998)							Third survey (2001–2003)									Fourth survey (2006–2008)								
	ND	S	B	SB	TN	D	P	ND	S	B	SB	TN	MR	PDD	D	P	ND	S	B	SB	TN	MR	PDD	D	P
Fraxinus platypoda	0	0	0	0	463	−1.00	<0.01	0	0	0	0	440	5.6	0	−1.00	<0.01	14	2	11	1	407	7.5	0	−0.75	<0.01
Cercidiphyllum japonicum	0	0	0	0	59	−1.00	<0.01	0	0	0	0	105	5.1	0	−1.00	<0.01	22	12	5	5	103	1.9	0	0.67	0.30
Pterocarya rhoifolia	0	0	0	0	118	−1.00	<0.01	0	0	0	0	108	9.3	0	−1.00	<0.01	12	8	2	2	92	14.8	6.3	0.52	<0.01
Acer shirasawanum	1	0	1	0	429	−0.98	<0.01	4	0	4	0	396	7.7	0	−0.90	<0.01	36	8	24	4	361	8.8	5.7	−0.32	<0.01
Acer pictum	0	0	0	0	272	−1.00	<0.01	0	0	0	0	257	6.3	0	−1.00	<0.01	4	3	1	0	242	5.8	0	−0.78	<0.01
Ulmus laciniata	3	0	3	0	95	−0.12	<0.01	15	1	14	0	84	17.9	17.6	0.71	<0.01	30	2	28	0	56	33.3	60.7	0.95	<0.01
Acer carpinifolium	0	0	0	0	510	−1.00	<0.01	119	118	0	1	586	5.9	16.7	−0.13	<0.01	343	264	12	67	529	9.7	17.5	0.68	<0.01
Carpinus cordata	0	0	0	0	53	−1.00	<0.01	0	0	0	0	51	7.5	0	−1.00	0.05	3	3	0	0	48	5.9	0	0.47	<0.01
Acer argutum	0	0	0	0	141	−1.00	<0.01	25	22	3	0	101	29.1	2.4	0.76	<0.01	39	12	10	17	52	48.5	42.9	0.98	<0.01
Pterostyrax hispida	0	0	0	0	89	−1.00	<0.01	3	3	0	0	67	30.3	3.7	0.23	0.63	14	13	0	1	40	40.3	33.3	0.93	0.21
Others	0	0	0	0	174			4	1	3	0	156					19	9	6	4	122				

ND total number of damaged trees, *S* feeding damage to leaves and stems, *B* debarking and fraying damage, *SB* both *S* and *B*, *TN* total number of standing trees, *MR* mortality rate (%), *PDD* percentage of browsed trees among all dieback trees, *D* Ivlev's modified electivity index (Jacobs 1974), *P* significance of the difference between the browsing–damage ratio for each species and the overall ratio (Fisher's exact test)

of 19 species, including some trees of the dominant canopy species (Table 9.3). The likelihood ratio test indicated that browsing damage had increased significantly throughout the survey period ($P < 0.001$), and that the extent of damage varied among species ($P < 0.001$). However, the interaction between year and species was not significant ($P = 0.96$).

9.5   Species and Size Preferences, and Mortality

Among the browsed tree species, *U. laciniata* and *A. argutum* had significantly higher ratios of damaged trees among the total number of trees in the forest ($P < 0.01$), and higher Ivlev's modified electivity indices (*D* values) in all surveys (Table 9.3). The index D_i (Jacobs 1974) for species i was calculated as follows:

$$D_i = \frac{r_i - P_i}{r_i + P_i - 2r_iP_i},$$

where r_i is the ratio of the number of damaged trees of the ith species to the total number of damaged trees, and P_i is the ratio of the total number of trees of the ith species to the total number of trees (Jacobs 1974). The index D is 1 for preferred species and -1 for non-preferred species. The D values for *P. hispida* and *A. carpinifolium* increased gradually throughout the survey period, reaching 0.93 and 0.68, respectively, by the fourth survey. By contrast, *F. platypoda* and *A. pictum* had lower proportions of damaged trees among the total number of trees ($P < 0.01$) and lower *D* values. Initially, browsing damage was found primarily on trees with smaller DBHs (4–20 cm); its distribution then expanded gradually to include larger trees (DBH > 20 cm; Fig. 9.2). The likelihood ratio test indicated that the DBH correlated negatively with browsing damage caused by the deer ($P < 0.01$). This correlation was particularly strong for *U. laciniata*, *A. argutum*, and *A. carpinifolium* (Fig. 9.3). Exceptionally, the DBH showed no obvious relationship to browsing damage to trees of *P. rhoifolia* and *C. japonicum*. During the tree census, we recorded browsing-damage levels and divided them into three categories: (1) feeding on leaves and stems; (2) debarking, including the fraying of bark; and (3) a combination of feeding and debarking damage (Fig. 9.4). Evidence of feeding on leaves and stems, debarking, and combined feeding and debarking damage was found on almost all dominant and abundant species in 2006–2008 (Table 9.3). Among the excessively browsed species, debarking damage occurred more frequently than feeding damage on *U. laciniata* and *A. shirasawanum*. By contrast, *A. carpinifolium* and *P. hispida* were subjected to heavier feeding on leaves and stems than on bark. Feeding and debarking damage were recorded in equal quantities on *A. argutum*. The dieback of damaged trees of excessively browsed species increased until the fourth survey (likelihood ratio test, $P < 0.01$; Fig. 9.3, Table 9.3). However, the mortality of damaged trees varied among species. For example, the mortality rates of

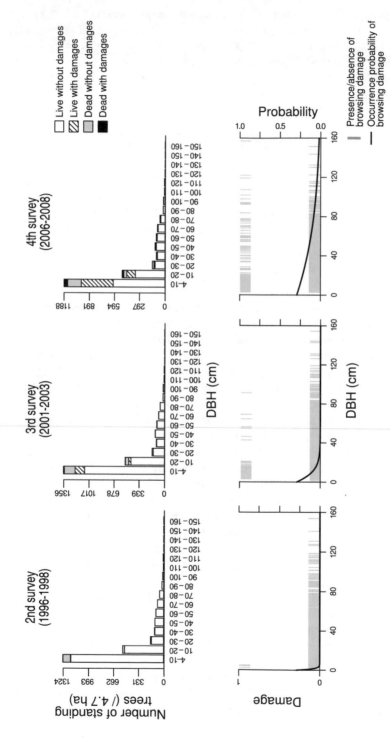

Fig. 9.2 Temporal changes in tree-size preference by sika deer and DBH distributions of all trees in the survey area of the Ooyamazawa riparian forest from the second through fourth surveys (upper three panels). The lower three panels indicate the relationship between DBH and the presence/absence of browsing damage, and estimated probabilities derived using generalized linear mixed models

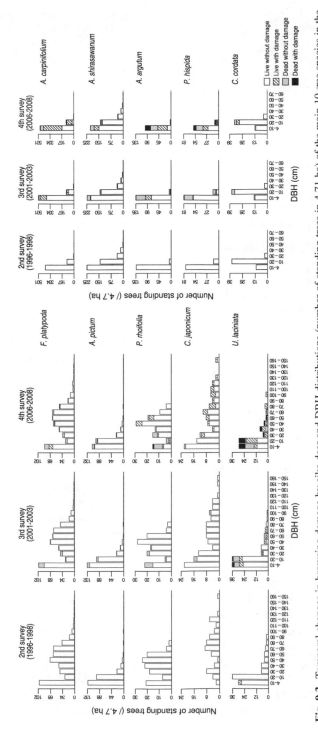

Fig. 9.3 Temporal changes in browsing damage by sika deer and DBH distributions (number of standing trees in 4.71 ha) of the main 10 tree species in the Ooyamazawa riparian forest from the second through fourth surveys. Browsing damage included feeding damage to leaves and stems, bark stripping, and fraying damage to bark

Fig. 9.4 Feeding on leaves and stems of *P. hispida* (left) and debarking of *A. carpinifolium* (right) by sika deer

U. laciniata, *A. argutum*, and *P. hispida* were high, whereas that of *A. carpinifolium* was relatively low (Table 9.3).

These results suggest that *A. argutum*, *U. laciniata*, *P. hispida*, and *A. carpinifolium* were preferentially browsed by deer compared with *F. platypoda*, *A. pictum*, and *A. shirasawanum* (Table 9.3). The diet of deer is determined by the physical ease of feeding, and the nutritional and lignin contents of plants (Gill 1992a; Jiang et al. 2005; Kojima et al. 2006). The major damage to *A. argutum*, *A. carpinifolium*, and *P. hispida* may have been caused by the relatively small size of these three species in the forest (Table 9.1). In general, DBH and tree height variation differ among species in natural forests. These three species have relatively low tree heights and small maximum DBHs, in addition to producing many sprouts without scarring caused by disturbances. Smaller trees have more leaves and stems at lower heights, accessible to deer. In addition, bark becomes thicker, harder, rougher, and more suberizing with increasing tree diameter. Therefore, these three species may be physically easy to feed on by deer. By contrast, *F. platypoda* and *A. pictum* have greater tree heights and trunk diameters. The forest-floor vegetation, which includes the seedlings and saplings of these two species, has almost disappeared due to heavy browsing pressure of deer, except for poisonous herbs such as *Veratrum album* subsp. *oxysepalum* (Turcz.) Hultén *Aconitum tonense* Nakai ex H.Hara, and *Scopolia japonica* Maxim. Thus, the lower preference for these species by deer, despite their palatability, may be caused by the presence of fewer leaves and stems at lower heights, in addition to thicker bark relative to other species. These results suggest that deer browsing activity especially reduces tree-species diversity of certain functional types, such as shrub species (i.e., species with relatively low tree heights and small maximum DBHs). On the other hand, *U. laciniata* was subject to the most damage among the four preferred species. The greater preference for this species has also been confirmed in previous studies (Imagawa and Tanaka 1996;

Kaji et al. 2010). This species may be targeted preferentially due to the low lignin content in its bark (Imagawa and Tanaka 1996; Kojima et al. 2006) and the physical ease of strip-peeling (Imagawa and Tanaka 1996).

9.6   Resistance of Trees to Deer Browsing

Heavy browsing, especially bark-stripping by large herbivorous mammals, induces the dieback of damaged trees (Gill 1992b; Akashi and Nakashizuka 1999). The increase in dieback of damaged trees among the four preferred species in our study (*A. argutum*, *U. laciniata*, *P. hispida*, and *A. carpinifolium*) during 2006–2008 suggests that deer browsing caused serious damage to tree species in this old-growth riparian forest (Fig. 9.3, Table 9.3). However, the mortality rates differed among the four species. *Acer argutum* and *U. laciniate*, which had high mortality rates, were subjected to more debarking than feeding damage. Conversely, *A. carpinifolium* had a lower mortality rate than did *A. argutum* and *U. laciniata*, even though 79 trees of *A. carpinifolium* were affected by debarking damage and 264 trees showed damage caused by feeding on stems and leaves (Table 9.3). These results indicate that these four species have different levels of resistance to deer browsing, with the highest being shown by *A. carpinifolium*. Similar findings were obtained in several previous studies based on simulated experiments of browsing and debarking (Noel 1970; Bergström and Danell 1987; Delvaux et al. 2010). The high resistance of *A. carpinifolium* is considered to be due to its high rates of adventitious bud production and regrowth of bark, including the cambium layer and phloem, after browsing and debarking.

9.7   Effective Management of Ooyamazawa

Previous studies have suggested that browsing by sika deer adversely affects the regeneration of tree species in natural forests (Takatsuki and Gorai 1994; Akashi and Nakashizuka 1999; Suda et al. 2001; Yokoyama et al. 2001). Our study reveals an increase in serious damage to tree species in an old-growth riparian forest due to 18 years of heavy browsing activity by deer. Riparian forests have important and unique ecosystem functions (Malanson 1993; Richards et al. 2002; Ward et al. 2002); however, the areas of these forests in mountain landscapes are smaller than those of other forests on mountain slopes, such as *Fagus crenata* forest. Therefore, deer browsing appears to negatively impact not only riparian forests, but also mountain forest ecosystems. If the deer population density is maintained at the current level or increases further, modification of the forest structure and composition will be induced by the failure of preferred species to regenerate and the dieback of trees with small diameters. To maintain the ecosystems of riparian forests, immediate preventive management is required before browsing damage by deer expands.

Deer-exclusion fences are an effective means of forest management and have been used widely throughout Japan. However, these fences can protect only limited areas of forest, as installation and maintenance costs are very high. The control of the deer population through hunting or culling is an alternative method, and the number of sika deer harvested has increased rapidly over the last few decades (Takatsuki 2009). However, population control by hunting and culling is expected to be insufficient because the number of hunters is gradually declining in Japan (Takatsuki 2009). Hence, temporal and immediate activities, such as the selective conservation of tree-seed sources, are needed until a viable management program becomes operational.

In the Ooyamazawa riparian forest, extensive browsing damage and increased tree mortality were found for *U. laciniata*, *A. argutum*, and *P. hispida*. Therefore, in other riparian forests of similar species composition and size structure, mature trees of these three species with small diameters (i.e., shrub species) should be primarily protected and managed through the use of metallic-mesh trunk protectors.

References

Akashi N, Nakashizuka T (1999) Effects of bark-stripping by sika deer (*Cervus nippon*) on population dynamics of a mixed forest in Japan. For Ecol Manage 113:75–82

Akashi N, Terazawa K (2005) Bark stripping damage to conifer plantations in relation to the abundance of sika deer in Hokkaido, Japan. For Ecol Manage 208:77–83

Ando M, Yokota H, Shibata E (2003) Bark stripping preference of sika deer, *Cervus nippon*, in terms of bark chemical content. For Ecol Manage 177:323–331

Augustine DJ, McNaughton SJ (1998) Ungulate effects on the functional species composition of plant communities: herbivore selectivity and plant tolerance. J Wildl Manage 62:1165–1183

Bergström R, Danell K (1987) Effects of simulated winter browsing by moose on morphology and biomass of two birch species. J Ecol 75:533–544

Boulangera V, Baltzingera C, Saïdc S, Ballona P, Picardb JF, Dupoueyb JL (2009) Ranking temperate woody species along a gradient of browsing by deer. For Ecol Manage 258:1397–1406

Côté SD, Rooney TP, Tremblay JP, Dussault C, Waller DM (2004) Ecological impacts of deer overabundance. Annu Rev Ecol Evol Syst 35:113–147

Delvaux C, Sinsin B, Damme PV (2010) Impact of season, stem diameter and intensity of debarking on survival and bark re-growth pattern of medical tree species, Benin, West Africa. Biol Conserv 143:2664–2671

Didion M, Kupferschmid AD, Bugmann H (2009) Long-term effects of ungulate browsing on forest composition and structure. For Ecol Manage 258S:S44–S55

Furusawa H, Miyanishi H, Kaneko S, Hino T (2003) Movement of soil and litter on the floor of a temperate mixed forest with an impoverished understory grazed by deer (*Cervus nippon centralis* temminck). J Jpn For Soc 85:318–325 (in Japanese with English summary)

Gill R (1992a) A review of damage by mammals in north temperate forests. 1. Deer. Forestry 65:145–169

Gill R (1992b) A review of damage by mammals in north temperate forests. 3. Impact on trees and forests. Forestry 65:363–388

Gunma, Saitama, Tokyo, Yamanashi and Nagano Prefectural Boards of Education (eds) (1988) Bulletin report of Japanese Serow (*Capricornis crispus*) protection area in Kanto mountains. Gunma, Saitama, Tokyo, Yamanashi and Nagano Prefectural Boards of Education, Tokyo (in Japanese)

Gunma, Saitama, Tokyo, Yamanashi and Nagano Prefectural Boards of Education (eds) (2010) Bulletin report of Japanese Serow (*Capricornis crispus*) protection area in Kanto mountains. Gunma, Saitama, Tokyo, Yamanashi and Nagano Prefectural Boards of Education, Tokyo (in Japanese)

Imagawa H, Tanaka K (1996) Damage of forest trees by sika deer. Possible reason for the palatability. For Tree Breed Hokkaido 39:10–13 (in Japanese)

Jacobs J (1974) Quantitative measurement of food selection. A modification of the forage ratio and Ivlev's electivity index. Oecologia 14:413–417

Jayasekara P, Takatsuki S (2000) Seasonal food habits of a sika deer population in the warm temperate forest of the westernmost part of Honshu, Japan. Ecol Res 15:153–157

Jiang Z, Ueda H, Kitahara M, Imaki H (2005) Bark stripping by sika deer on veitch fir related to stand age, bark nutrition, and season in northern mount Fuji district, central Japan. J For Res 10:359–365

Kaji K, Miyaki M, Saitoh T, Ono S, Kaneko M (2000) Spatial distribution of an expanding sika deer population on Hokkaido Island, Japan. Wildl Soc Bull 28:699–707

Kaji K, Saitoh T, Uno H, Matsuda H, Yamamura K (2010) Adaptive management of sika deer populations in Hokkaido, Japan: theory and practice. Popul Ecol 52:373–387

Kanda N, Yokota T, Shibata E, Sato H (2005) Diversity of dung-beetle community in declining Japanese subalpine forest caused by an increasing sika deer population. Ecol Res 20:135–151

Kojima Y, Yasui Y, Orihashi K, Terazawa M, Kamoda S, Kasahara H, Takahashi Y (2006) Relationship between bark preferences of *Cervus nippon yesoensis* and inner bark components of small-diameter tree trunk. J Jpn For Soc 88:337–341 (in Japanese with English summary)

Malanson G (1993) Riparian landscapes. Cambridge University Press, Cambridge

Maruyama N, Furubayashi K (1983) Preliminary examination of block count method for estimating number of sika deer in Fudakake. J Mamm Soc J 9:274–278 (in English with Japanese summary)

Moser B, Schütz M, Hindenlang KE (2006) Importance of alternative food resources for browsing by roe deer on deciduous trees: the role of food availability and species quality. For Ecol Manage 226:248–255

Niwa S, Mariani L, Kaneko N, Okada H, Sakamoto K (2011) Early-stage impacts of sika deer on structure and function of the soil microbial food webs in a temperate forest: a large-scale experiment. For Ecol Manage 261:391–399

Noel A (1970) The girdled tree. Bot Rev 36:162–195

Nomiya H, Suzuki W, Kanazashi T, Shibata M, Tanaka H, Nakashizuka T (2003) The response of forest floor vegetation and tree regeneration to deer exclusion and disturbance in a riparian deciduous forest, central Japan. Plant Ecol 164:263–276

Oi T, Suzuki M (2001) Damage to sugi (*Cryptomeria japonica*) plantations by sika deer (*Cervus nippon*) in northern Honshu, Japan. Mammal Study 26:9–15

Richards K, Brasington J, Hughes F (2002) Geomorphic dynamics of floodplains: ecological implications and a potential modelling strategy. Freshwat Biol 47:559–579

Ripple WJ, Beschta RL (2008) Trophic cascades involving cougar, mule deer, and black oaks in Yosemite National Park. Biol Conserv 141:1249–1256

Saitama Museum of Natural History (eds) (1983) Bulletin report of special natural monument: density of Japanese Serow. Saitama Prefectural Boards of Education (in Japanese)

Sakio H (1997) Effects of natural disturbance on the regeneration of riparian forests in a Chichibu mountains, central Japan. Plant Ecol 132:181–195

Sakio H (2008) Features of riparian forests in Japan. In: Sakio H, Tamura T (eds) Ecology of riparian forests in Japan: disturbance life history and regeneration. Springer, Tokyo, pp 3–11

Sakio H, Kubo M, Kawanishi M, Higa M (2013) Effects of deer on forest floor vegetation in the Chichibu mountains, Japan, J Jpn Soc Reveg Technol 39:226–231 (in Japanese with English summary)

Salk TT, Frelich LE, Sugita S, Calcote R, Ferrari JB, Montgomery RA (2011) Poor recruitment is changing the structure and species composition of an old-growth hemlock-hardwood forest. For Ecol Manage 261:1998–2006

Sauvé DG, Côté SD (2006) Winter frage selection in white-tailed deer at high density: balsam fir is the best of a bad choice. J Wildl Manage 71:911–914

Suda K, Araki R, Maruyama N (2001) The effects of sika deer on the structure and composition of the forests on the Tsushima islands. Biosphere Conserv 4:13–22

Suzuki M, Miyashita T, Kabaya H, Ochiai K, Asada M, Tange T (2008) Deer density affects ground-layer vegetation differently in conifer plantations and hardwood forests on the Boso peninsula, Japan. Ecol Res 23:151–158

Takahashi H, Kaji K (2001) Fallen leaves and unpalatable plants as alternative foods for sika deer under food limitation. Ecol Res 16:257–262

Takatsuki S (2009) Effects of sika deer on vegetation in Japan: A review. Biol Conserv 142:1922–1929

Takatsuki S, Gorai T (1994) Effects of sika deer on the regeneration of a *Fagus crenata* forest on Kinkazan Island, northern Japan. Ecol Res 9:115–120

Ueda A, Hino T, Ito H (2009) Relationships between browsing on dwarf bamboo (*Sasa nipponica*) by sika deer and the structure of ground beetle (coleoptera: Carabidae) assemblage. J Jpn For Soc 91:111–119 (in Japanese with English summary)

Ueda H, Takatsuki S, Takahashi Y (2002) Bark stripping of hinoki cypress by sika deer in relation to snow cover and food availability on Mt Takahara, central Japan. Ecol Res 17:545–551

Uno H, Yokoyama M, Sakata H, the Working Group for Sika Deer Management Mammalogical Society of Japan (2007) Current status of and perspectives on conservation and management for sika deer populations in Japan. Mamm Sci 47:25–38 (in Japanese with English summary)

Ward JV, Tockner K, Arscott DB, Claret C (2002) Riverine landscape diversity. Freshwat Biol 47:517–539

Yokoyama S, Maeji I, Ueda T, Ando M, Shibata E (2001) Impact of bark stripping by sika deer, *Cervus nippon*, on subalpine coniferous forests in central Japan. For Ecol Manage 140:93–99

Chapter 10
Characteristics and Temporal Trends of a Ground Beetle (Coleoptera: Carabidae) Community in Ooyamazawa Riparian Forest

Shigeru Niwa

Abstract Over the course of a 10-year monitoring survey of ground-dwelling beetles in the Ooyamazawa riparian forest (2008–2017), 2381 individuals from 19 beetle families (including 1969 individuals of 36 carabid species) were collected. The carabid community was characterized by high species richness and a high proportion of endemic species, when compared with other monitored forest sites in Japan. Most of the carabid species exhibited drastic declines in abundance, and the annual catch of the carabids over the 10-year period decreased by 80%. These declines of the beetles were likely to be caused by changes in the forest floor environment due to overabundance of deer as well as climate warming.

Keywords Global warming · Long-term ecological research · Monitoring Sites 1000 Project · Pitfall trap · Sika deer overabundance

10.1 Introduction

The ground beetles (i.e., carabids; Coleoptera: Carabidae) comprise a species-rich family that contains over 40,000 described species. Carabids are globally distributed and are found in a wide range of terrestrial ecosystems (e.g., forest, grassland, wetland, seashore, desert, tundra, alpine zone, cave, arable land, and urban green area). Individuals live for one or several years and usually reproduce on an annual basis (Lövei and Sunderland 1996). Carabids that inhabit temperate forests are often flightless, live on the ground, and mainly prey on ground and soil invertebrates (Shibuya et al. 2014, 2018; Okuzaki et al. 2009; Vanbergen et al. 2010), while also serving as prey for a variety of vertebrates (Lövei and Sunderland 1996). The low dispersal ability of forest-dwelling carabids results in high rates of local endemism,

S. Niwa (✉)
Japan Wildlife Research Center, Tokyo, Japan
e-mail: sniwa@jwrc.or.jp

© The Author(s) 2020
H. Sakio (ed.), *Long-Term Ecosystem Changes in Riparian Forests*, Ecological Research Monographs, https://doi.org/10.1007/978-981-15-3009-8_10

179

as well as significant differences in the structure of geographically isolated communities, and may also increase the susceptibility of such species and communities to environmental changes.

The species composition of carabid communities largely varies by habitat, owing to differences in species' life histories (e.g., seasonality), dispersal abilities (e.g., flight ability), feeding habits (e.g., carnivore, granivore, or omnivore), habitat and microenvironment preferences (e.g., vegetation structure, shadiness, temperature, moisture, soil particle size, and hibernation site; Lövei and Sunderland 1996; Thiele 1977). The species composition is also likely affected by landscape structure, owing to the beetles' low dispersal ability, and is, therefore, likely sensitive to both natural and anthropogenic disturbances, including fire, grazing, grassland and forest management, agricultural practice, pollution, and habitat fragmentation (Rainio and Niemelä 2003; Koivula 2011). In addition, sufficiently large and quantitative carabid samples can be easily and cost-effectively collected by pitfall-trapping. For these features, carabid communities are often used as bioindicators (Rainio and Niemelä 2003; Koivula 2011) and have recently been adopted as a monitoring target for broad-scale, long-term ecological monitoring programs, including the Environmental Change Network (ECN) in the UK (Morecroft et al. 2009), the National Ecological Observatory Network (NEON) in the USA (Hoekman et al. 2017), and the Monitoring Sites 1000 Project in Japan (Niwa et al. 2016). Indeed, the data collected by the ECN from 1994 to 2011 allowed researchers to document substantial declines in carabid biodiversity throughout the UK and significant phenological shifts in many carabid species in Scotland (Brooks et al. 2012; Pozsgai and Littlewood 2014).

The Ooyamazawa riparian forest (Chichibu Mountains, Japan) was designated as a core site of the Forest and Grassland Survey in the Monitoring Sites 1000 Project, which is a nationwide, long-term ecological monitoring project that was launched by the Ministry of Environment (Japan) in 2003, and tree, bird, and ground-dwelling beetle surveys have been conducted at this site on an annual basis since 2008. For the Monitoring Sites 1000 Project, ground-dwelling beetle surveys have been performed in 22 forests across Japan (Niwa et al. 2016), and because most of these forests have not been subjected to direct human disturbance (e.g., logging) for over 100 years, the resulting data can be used to study natural gradients of beetle diversity within the Japanese Archipelago, as well as the effects of broad-scale environmental changes (e.g., climate change). The survey protocols and most of the data obtained in this project are publicly available *via* the internet (http://www.biodic.go.jp/moni1000/index.html, Ishihara et al. 2011; Suzuki et al. 2012; Niwa et al. 2016).

The Ooyamazawa riparian forest is a well-preserved old-growth forest and that possesses a diverse forest floor environment provided by the stream. Such a stable and continuous forest ecosystem with a diverse forest floor environment can be expected to harbor a rich and diverse carabid beetle community. Recently, however, sika deer populations have increased and the resulting heavy browsing pressure has completely changed the vegetation of the forest floor (Chap. 8). Indeed, the alteration of forest vegetation by deer overabundance has been reported all over Japan (Takatsuki 2009), as well as in subarctic and temperate zones all over the world, and the cascading effects of such population growth on other animals and ecosystem

Fig. 10.1 Annual variation in mean air temperature at Chichibu weather station. Data were obtained from the Japan Meteorological Agency (http://www.data. jma.go.jp/obd/stats/etrn/ index.php). Dotted line indicates the 10-year moving average

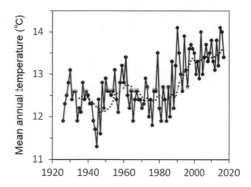

processes have become serious issues (Côté et al. 2004; Rooney and Waller 2003; Stewart 2001; Foster et al. 2014). In recent decades, the Ooyamazawa forest has also been affected by temperature increases associated with global warming. The mean air temperature in Chichibu City, for example, which is located 28 km east of Ooyamazawa, has increased at ~1.5 °C per 100 years, and the last decade was the hottest decade of the last 90 years in Chichibu (Fig. 10.1).

In this chapter, I will describe the diversity and characteristics of the carabid beetle community in the Ooyamazawa riparian forest and describe how the community has changed during the initial 10 years (2008–2017) of the long-term monitoring project.

10.2   Study Site and Methods

In the Forest and Grassland Survey in the Monitoring Sites 1000 Project, ground-dwelling beetle surveys have been conducted at 22 forest sites, including Ooyamazawa, every year since 2004, 2005, 2006, or 2008 (Niwa et al. 2016). Each of the 22 forest sites is primarily composed of natural old-growth forests, and the sites span both the major climatic zones (subalpine, cool-temperate, warm-temperate, and subtropical) and forest types (coniferous, deciduous broadleaved, evergreen broadleaved, and conifer and broadleaf mixed) of Japan. At each site, one monitoring plot (100 m × 100 m) and five subplots (5 m × 5 m) within it have been established, and four pitfall traps (9-cm diameter, 12-cm depth, with no baits or preservatives) were installed in each subplot (Fig. 10.2). The traps were opened for 3 d each in four seasons per year (spring: May–June, summer: June–July, early-autumn: September–October, and late-autumn: October–November), and the animals captured in the traps during each sampling period were collected, identified to class, order, or family level, enumerated, dried, weighed, and preserved as dry specimens. Captured beetles were further identified to family, genus, or species level and individually weighed. The species were identified following the taxonomic system of Löbl and Löbl (2017).

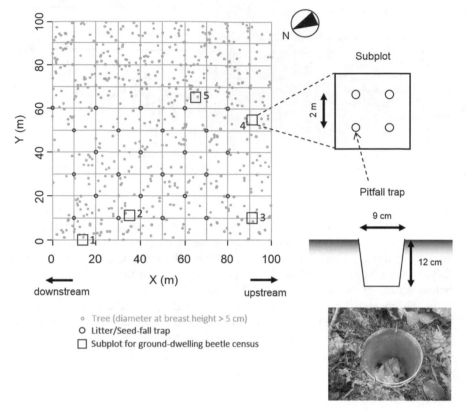

Fig. 10.2 Study plot established for the long-term monitoring of trees, litter and seed fall, and ground-dwelling beetles in the Ooyamazawa riparian forest

In Ooyamazawa, beetle sampling has been conducted in May, June, September, and October of every year since 2008. However, data from 2013 and 2015 have been excluded from the following analyses because the access road was temporarily impassable, owing to typhoon damage, and sampling could not be conducted in September of either year.

The general description of the monitoring plot in Ooyamazawa, including climate, geology, topography, soil and vegetation, is provided in Chap. 1. Subplots 1 and 2 were located on the bottom of a valley, along with a small first-order stream that usually flows belowground, and the colluvial deposits (gravels of various sizes) in these subplots were covered by soil, with a layer of litter. Meanwhile, Subplots 3 and 4 were located at the lower end of the slopes, and because the rich forest floor vegetation (e.g., ferns in Subplots 1–4 and dwarf bamboo in Subplot 4) had declined before the initiation of monitoring (Chap. 8), scarce vegetation was present in the subplots throughout the monitoring period (2008–2017). In contrast, Subplot 5 located in the middle of a slope had a relatively deep soil layer, and was, at least initially, covered with dense forest floor vegetation of dwarf bamboo (*Sasa borealis*). The bamboo vegetation, however, rapidly declined and almost disappeared during the monitoring period (Chap. 8).

10.3 Carabid Community Diversity and Distinctness

From 2008 to 2017, 2381 beetles from 19 families were collected (Table 10.1). Carabid beetles (36 species) accounted for 82.7% of them, and three predominant genera (*Carabus*, *Pterostichus*, and *Synuchus*) accounted for 92.5% of the carabids (Table 10.2, Fig. 10.3). The carabid community of Ooyamazawa was characterized by high species diversity, in comparison to the 21 other monitoring sites. The mean annual species number (17.8 ± 1.0, mean ± standard error) and diversity index (Shannon-Wiener's H': 2.19 ± 0.07) were the highest among these forests, whereas the mean annual catch (132.3 ± 32.5) was relatively high (sixth highest) and the evenness (Pielou's J': 0.768 ± 0.032) was about average.

The high species diversity of the carabids in Ooyamazawa can be partly attributed to the stable forest environment and to the presence of a stream, owing to higher soil moisture, greater habitat heterogeneity (including sand or gravel bars), a distinctive plant community, aquatic food resources, and occasional disturbance by flooding.

Table 10.1 Total catch of beetle families in the Ooyamazawa riparian forest, during 2008–2017

Family	Individuals
Carabidae	1969
Hydrophilidae	4
Ptiliidae	1
Leiodidae	66
Scydmaenidae	3
Silphidae	32
Staphylinidae	138
Geotrupidae	66
Lucanidae	8
Scarabaeidae	9
Nitidulidae	1
Cryptophagidae	5
Endomychidae	1
Corylophidae	1
Tetratomidae	1
Tenebrionidae	3
Pyrochroidae	1
Chrysomelidae	1
Curculionidae	7
Unknown (adult)	1
Unknown (larva)	63
Total	2381

Table 10.2 Annual catch and traits of each carabid species captured in the Ooyamazawa riparian forest, during 2008–2017

Species	Annual catch[a] (indiv./20 traps·12 days)	Distribution[b]	Body length[c] (mm)	Hind wing[d]	Adult active season[e]	Note
Nebria sadona	1.63 (0.60)	HON, SHI, KYU, Sad	11.0–14.5	B	A	N. chichibuensis?
Notiophilus impressifrons	0.13 (0.13)	HOK, HON, SHI, KYU, Sak; AS, EU, NA	5.7–7.0	M	S	
Carabus albrechti	29.25 (8.20)	HOK, HON, Sad	19.0–24.0	B	S	ssp. esakianus
Carabus procerulus	5.25 (0.92)	HON, KYU	25.0–35.0	B	A	ssp. procerulus
Carabus arboreus	5.25 (1.38)	HOK, HON, Sak	18.0–25.0	B	A	ssp. ogurai
Carabus harmandi	9.13 (2.66)	HON	18.0–23.0	B	S	ssp. okutamaensis
Brachinus stenoderus	0.25 (0.17)	HOK, HON, SHI, KYU; AS	5.5–11.5	M	S	
Trichotichnus lewisi	0.13 (0.13)	HOK, HON, SHI, KYU	12.5–14.5	M	A	
Trichotichnus yoshiroi	1.50 (0.35)	HON (Kanto Mts.)	9.4–10.1	B	S	
Trichotichnus sp.	0.13 (0.13)	–	9.0	–	–	
Negreum bentonis	4.63 (2.43)	HON	9.0–12.0	B	S	
Platynus subovatus	0.13 (0.13)	HON	11.0–13.5	B	S	
Xestagonus xestum	0.25 (0.17)	HON, SHI, KYU; AS	8.0–11.0	B	S	
Myas cuprescens	2.88 (1.17)	HON, SHI, KYU, Sad	14.0–22.5	B	A	ssp. cuprescens
Pterostichus subovatus	5.63 (2.44)	HOK, HON, KYU; AS	11.0–14.5	M	S	
Pterostichus karasawai	0.13 (0.13)	HON	13.5–16.0	B	S	
Pterostichus koheii	0.13 (0.13)	HON (Kanto Mts.)	15.0–16.0	B	–	
Pterostichus mucronatus	0.63 (0.28)	HON	16.5–22.0	B	S	
Pterostichus yoritomus	0.50 (0.29)	HON, KYU	12.0–14.5	B	S	
Pterostichus katashinensis	3.13 (0.77)	HON	15.5–19.0	B	S	ssp. naganoensis
Pterostichus mitoyamanus	1.88 (0.77)	HON (Kanto Mts.)	15.5–18.0	B	S	
Pterostichus okutamae	1.25 (0.34)	HON (Kanto Mts.)	15.5–17.5	B	S	
Pterostichus rhanis	2.50 (0.95)	HON	13.5–16.0	B	S	ssp. kantous
Pterostichus tokejii	9.88 (2.36)	HON (Kanto Mts.)	17.0–19.5	B	S	

Pterostichus brittoni	0.25 (0.17)	HON	8.2–8.8	B	S	
Stomis prognathus	0.25 (0.17)	HON, SHI	9.0–12.0	B	S	
Pristosia aeneola	0.13 (0.13)	HON, SHI	11.5–16.0	B	A	
Parabroscus crassipalpis	–	HOK, HON, SHI, KYU, Izu	11.0–13.0	M	A	
Synuchus agonus	3.75 (1.19)	HON, SHI, KYU, Kur; AS	9.0–10.0	B	A	
Synuchus arcuaticollis	0.38 (0.28)	HOK, HON, SHI, KYU, Kur, Yak; AS	8.0–10.5	B/M	A	
Synuchus atricolor	0.13 (0.13)	HON; AS	11.0–15.0	B	A	
Synuchus crocatus	–	HOK, HON, SHI, KYU, Kur; AS	9.0–12.0	B/M	A	
Synuchus cycloderus	0.25 (0.27)	HOK, HON, SHI, KYU; AS	10.5–14.0	M	A	
Synuchus melantho	40.50 (17.05)	HOK, HON, SHI, KYU, Kur; AS	9.5–13.0	B/M	A	
Synuchus nitidus	–	HOK, HON, SHI, KYU, Kur, Tai; AS	12.5–17.0	M	A	
Synuchus tanzawanus	0.38 (0.20)	HON	7.4–9.0	B	A	
Synuchus sp.[f]	0.13 (0.13)	–	–	–	–	
Carabidae Gen. sp.[f]	0.13 (0.13)	–	–	–	–	

[a]Total of four seasons and five subplots. Average and standard error (in parentheses) from 2008 to 2017 (excluding 2013 and 2015). –, captured only in 2013 and/or 2015

[b]Main islands and continents (Ueno et al. 1985; Löbl and Löbl 2017; Imura and Mizusawa 2013; Habu 1978; Morita 1997): HOK Hokkaido, HON Honshu, SHI Shikoku, KYU Kyushu, Kur Kuril Isls., Sak Sakhalin, Sad Sado, Izu Izu Isls., Tsu Tsushima, Yak Yaku, Tai Taiwan, AS Asian continent, EU European continent, NA North American continent

[c]Ueno et al. (1985), Imura and Mizusawa (2013), Habu (1978), and Morita (1997)

[d]B degenerated (brachypterous), M developed (macropterous), B/M dimorphic

[e]S Spring–Summer, A Autumn

[f]Unidentifiable, due to damage

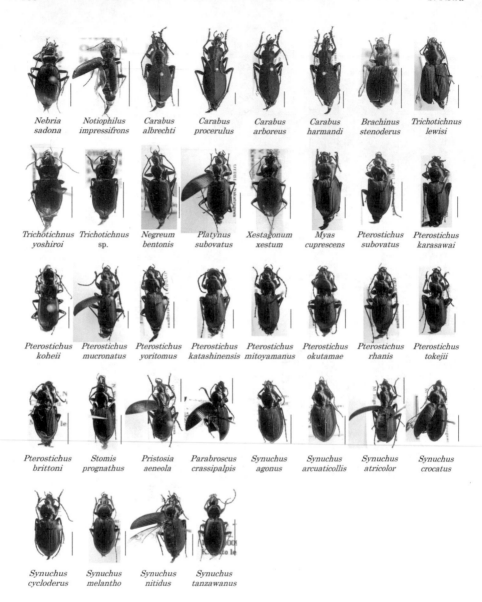

Fig. 10.3 Carabid species found in the Ooyamazawa riparian forest. Scale bars indicate 5 mm

For example, some of the species captured at Ooyamazawa (*Nebria sadona* and *Brachinus stenoderus*) have been reported to favor riverine gravel bars (Ueno et al. 1985; Terui et al. 2017). Multivariate analysis revealed that adjacent subplots possessed similar species compositions, which indicated the contribution of small-scale habitat heterogeneity to plot-level diversity (Fig. 10.4). In particular, the Autumn-breeding and more widely distributed species (i.e., *Carabus procerulus*,

Fig. 10.4 Spatial variation in the species composition of carabids captured in the Ooyamazawa riparian forest. Data were ordinated using principal component analysis (PCA), with partialing out the effect of year, and rare species (<10 individuals) were excluded from analysis. (**a**) Mean and standard error of site scores among years in each subplot. (**b**) Species score for each carabid species. Species names are abbreviated by letters: NbSa, *Nebria sadona*; CaAl, *Carabus albrechti*; CaPr, *Carabus procerulus*; CaAr, *Carabus arboreus*; CaHa, *Carabus harmandi*; TrYo, *Trichotichnus yoshiroi*; NgBe, *Negreum bentonis*; MyCu, *Myas cuprescens*; PtSu, *Pterostichus subovatus*; PtKt, *Pterostichus katashinensis*; PtMi, *Pterostichus mitoyamanus*; PtOk, *Pterostichus okutamae*; PtRh, *Pterostichus rhanis*; PtTo, *Pterostichus tokejii*; SyAg, *Synuchus agonus*; SyMe, *Synuchus melantho*

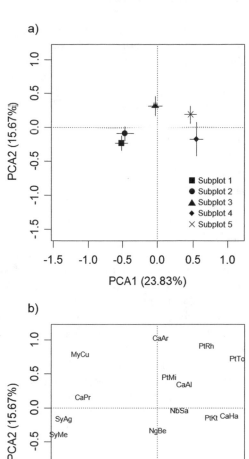

Myas cuprescens, and *Synuchus* spp.) were dominant in the subplots close to the streambed (Subplots 1 and 2), where the ground was covered with colluvial deposits and where ferns used to dominate. In contrast, the Spring- and Summer-breeding Honshu-endemic species (*Carabus harmandi* and *Pterostichus* (*Nialoe*) spp.) characterized the beetle communities of subplots on the slope with dwarf bamboo (Subplots 4 and 5), which may reflect the beetles' preference for moist conditions, habitat stability, or both. Furthermore, the fact that Ooyamazawa is located in a vast and continuous forested area that ranges from lowland areas to alpine zones in central Honshu Island and that includes a variety of forest types may also contribute to the site's high beetle diversity. Indeed, the Ooyamazawa forest is located in the middle altitude zone of this huge species pool of forest carabids and harbors species that are often found in both subalpine to alpine zones (e.g., *N. sadona*, *Carabus arboreus*, *C. harmandi*, *Xestagonum xestum*, and *Pristosia aeneola*) and in lowland forests (e.g., *Carabus albrechti*, *C. procerulus*, *Pterostichus yoritomus*, *Synuchus arcuaticollis*, *S. cycloderus*, and *S. nitidus*).

Another distinctive feature of the carabid species composition in Ooyamazawa forest is the high proportion of Japan-endemic species (64.9%) which was third highest among the 22 monitored forest sites. In particular, the site harbored many locally endemic species and subspecies limited to the Kanto Mountains and the surrounding area. For example, the *C.* (*Ohomopterus*), *C.* (*Leptocarabus*), *C.* (*Pentacarabus*) *harmandi*, *P.* (*Nialoe*), *P. macrogenys*, and *Trichotichnus leptopus* groups mainly composed of forest-inhabiting and flightless species and that are finely differentiated on Honshu Island (Imura and Mizusawa 2013; Morita 1997; Sasakawa 2009; Sasakawa and Kubota 2009; Sota and Nagata 2008; Zhang and Sota 2007). Among the species and subspecies found in Ooyamazawa, *C.* (*L.*) *arboreus ogurai*, *C.* (*P.*) *harmandi okutamaensis*, *P.* (*N.*) *tokejii*, *P.* (*N.*) *mitoyamanus*, *P.* (*N.*) *okutamae*, *P.* (*N.*) *rhanis kantous*, *P. koheii* (*P. macrogenys* group), and *T. yoshiroi* (*T. leptopus* group) are endemic to the Kanto Mountains, and *C.* (*O.*) *albrechti esakianus*, *P. karasawai*, *P. mucronatus*, *P.* (*N.*) *katashinensis naganoensis*, *P. brittoni*, and *S. tanzawanus* are endemic to relatively narrow region (mainly the Kanto and Chubu districts) that includes the Kanto Mountains (Habu 1958, 1978; Imura and Mizusawa 2013; Morita 1997; Sasakawa 2005a, b, 2009; Ueno et al. 1985). Sasakawa (2005a) investigated the phylogenetic relationships within the *P.* (*N.*) *asymmetricus* species group, which includes *P. tokejii*, *mitoyamanus*, *okutamae*, *rhanis*, and *katashinensis*, and suggested that the Kanto Mountains are a hotspot for speciation in this group. Furthermore, recent studies of *Nebria* (*Sadonebria*) have revealed that the group has also finely differentiated within the Honshu, Shikoku, and Kyushu Islands (Sasakawa 2016). Because it is possible that the *N.* (*S.*) *sadona* recorded in our site is actually *N.* (*S.*) *chichibuensis*, which could be an endemic of the Kanto Mountains, further investigation is needed for this species.

The low proportion of non-endemic continental species in the cool-temperate forests of Honshu Island, including Ooyamazawa, may be partly explained by the geohistorical background of the Japanese Archipelago. In particular, the history of land bridge formation between the Japanese Islands and the Asian continent can be an important factor affecting the distribution patterns of flightless animal species. Studies of paleoceanography and fossil records have revealed that southwestern Japan was often connected to the Asian continent by land bridges from the late Miocene to the Pleistocene (Tada 1994) and that land mammal species repeatedly immigrated to Japan using these bridges (Dobson and Kawamura 1998). During the late Pleistocene or last glacial period, Hokkaido was often connected to the Asian continent, and many land mammal species immigrated to Hokkaido during those periods, but most of those species did not immigrate to Honshu, which suggests that the connection between Hokkaido and Honshu, during this period, was temporal and unstable (Kawamura 2007). Dobson and Kawamura (1998) discussed that the terrestrial mammals that colonized Japan earlier than the middle Pleistocene have already speciated into Japan-endemic species, whereas some of the species that immigrated during later periods (middle to late Pleistocene) have not yet differentiated and remain non-endemic species. A phylogeographic study of *Carabus* (*Ohomopterus*), which is a representative group of the Japan-endemic carabids, showed that the ancestor of this group started to differentiate from its continental sister group (*C.* (*Isiocarabus*)) during the early Pleistocene, when stable connection between southwestern Japan and the Asian continent was broken, and then speciated rapidly in the Honshu, Shikoku, and Kyushu Islands (Sota and Nagata 2008). In a similar manner, Japan-endemic members of *C.* (*Leptocarabus*) differentiated from continental sister species and diverged into multiple endemic species when expanding eastward (Zhang and Sota 2007). In comparison to these endemic groups, non-endemic *Carabus* spp. exhibit smaller genetic differences with continental populations, which implies that they colonized Japan more recently using the southwestern or northern land bridges (Tominaga et al. 2000). For the continental carabid species that are adapted to warm-temperate climates and that have invaded Honshu *via* southwestern land bridges, the colonization of cool-temperate forests could be difficult. On the other hand, the dispersal from Hokkaido to Honshu by cold-adapted continental species that invaded Hokkaido using the northern land bridges during the last glacial period could be limited by the deep Tsugaru Strait, as has been reported for terrestrial mammals. Therefore, such geohistorical situations might have contributed to the current low proportion of non-endemic carabids in the cool-temperate forests of Honshu, including Ooyamazawa.

10.4   Trends in Carabid Abundance and Biomass

The annual abundance and biomass of carabid beetles decreased drastically over the last 10 years, by up to one-fifth and one-seventh, respectively (Fig. 10.5a, b), and these declines were the most prominent decreasing trends observed among the 22 forest monitoring sites. Furthermore, eight of the 12 dominant species (≥ 20 individuals were caught over 10 years) exhibited significant decreasing trends (Fig. 10.6), and the overall richness decreases, as well. However, the diversity index was not reduced, owing to increases in evenness that mainly resulted from large reductions in predominant species (e.g., *Synuchus melantho*, *Carabus albrechti*; Fig. 10.5d–f).

Over the last two decades, the overabundance of sika deer has caused rapid reductions in the forest floor vegetation of Ooyamazawa (Chap. 8). For example, the coverage of forest floor vegetation was >80% in 1998 and was reduced to only 3% by 2004 (Sakio et al. 2013). The vegetation cover in the five subplots, which was measured during each beetle survey (2008–2017), has remained low (5–30% in June). Reductions in forest floor vegetation by mammalian herbivores reduce litter fall, destabilize the litter layer and surface soil, promote the loss of litter layer, and cause the erosion of soil from steep slopes. In addition, trampling by the mammals

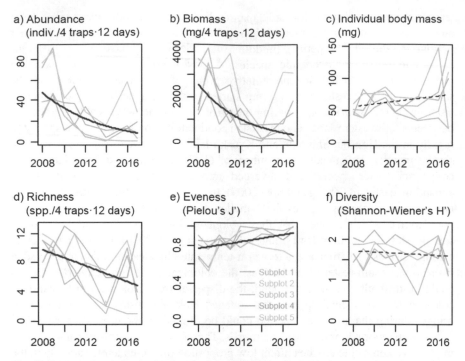

Fig. 10.5 Annual variation in the abundance, total biomass, individual mass, and species diversity of carabids in the Ooyamazawa riparian forest. Solid and dashed lines indicate significant and insignificant regressions by year, respectively (likelihood ratio test, $\alpha = 0.05$). Red and blue lines indicate increasing and decreasing trends, respectively. (**a**) Carabid abundance, Poisson regression; (**b**) Total carabid biomass, linear regression after log-transformation; (**c**) Individual body mass (biomass/abundance), linear regression; (**d–f**) Species diversity, linear regression. "Subplot" was used as a random factor

and the loss of forest floor cover, which protects soil from raindrop impact, are likely to promote the hardening of the soil surface and to reduce both the water permeability and water holding capacity of the forest soil. Loss of vegetation cover also reduced humidity, increases temperature, increases the variability of humidity and temperature at ground level, and reduces the availability of refuges for prey species to escape predators (e.g., mammals, birds). As a result, the changes brought by herbivore overabundance can generally reduce food resources and degrade the habitat of ground- and soil-dwelling animals, thereby reducing species density and diversity (e.g., Bressette et al. 2012; Lessard et al. 2012; Suominen 1999). In Ooyamazawa, the observed declines in carabid abundance, biomass, and richness could be caused directly by changes in the forest floor environment or indirectly by decreases in ground- and soil-dwelling invertebrates that serve as prey. The reduction of carabids occurred in all the subplots but was most conspicuous in Subplot 5, where the annual catch was reduced to almost zero (1 or 2 individuals per year) by 2014 (Fig. 10.5a). In this subplot, the dense cover of dwarf bamboo declined rapidly during the monitoring period. In addition, the beetles that initially inhabited Subplot

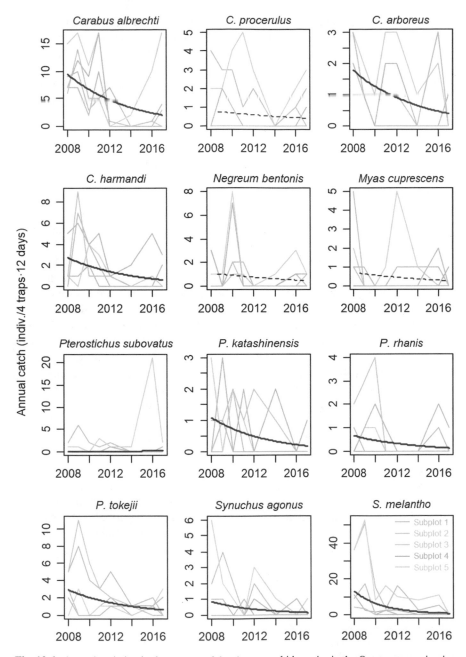

Fig. 10.6 Annual variation in the capture of dominant carabid species in the Ooyamazawa riparian forest. Dominant species were defined as those for which ≥20 individuals were captured over 10 years. Solid and dashed lines indicate significant and insignificant Poisson regressions by year, respectively (likelihood ratio test, $\alpha = 0.05$). Red and blue lines indicate increasing and decreasing trends, respectively. "Subplot" was used as a random factor

5, which were characterized by a high proportion of local endemics (Fig. 10.4), could have been most susceptible to deer overbrowsing because the subplot is located on a steep slope and because the litter layer and surface soil of the subplot are more likely to move without the support of forest floor vegetation.

However, the responses of ground- and soil-dwelling animals to increasing populations of large mammal herbivore are not always negative and vary, depending on taxon, ecological characteristics (e.g., body size, feeding habit, and habitat preference), environment (e.g., biome and productivity), and herbivory intensity (Bardgett and Wardle 2003; Stewart 2001; Foster et al. 2014; Allombert et al. 2005; Wardle et al. 2001; Flowerdew and Ellwood 2001). In the case of carabid beetles, the response of total abundance was highly variable because species that prefer open, warm, or dry conditions can be positively affected by the effects of large herbivores (Stewart 2001). For example, carabid beetles populations are reportedly increased by ungulate browsing in boreal forests, possibly owing to temperature increases and humidity decreases at the forest floor (Melis et al. 2006, 2007; Suominen and Danell 1999; Suominen et al. 2003). On the contrary, in Japanese temperate forests, many studies have reported that populations of certain small carabids increase in response to deer browsing, whereas populations of most large carabids (*Carabus* spp.) tend to decrease, and that, as a result, the dominancy of small species increased (Ueda et al. 2009; Okada and Suda 2012; Sato et al. 2018; Takakuwa et al. 2007; Yamada and Takatsuki 2015). In the Ooyamazawa forest, however, most species exhibited decreasing trends, regardless of body size, and only one small species (*Pterostichus subovatus*) exhibited a slight increasing trend (Figs. 10.6 and 10.7). Consequently, neither reductions in mean body mass nor directional changes in species composition were detected during the monitoring period (Figs. 10.5c and 10.8), possibly because few of the species at this site responded positively to deer browsing. Among the 12 dominant species, only one species (*P. subovatus*) had been reported by previous studies to respond positively to deer browsing, whereas negative responses had been reported for *C. procelurus*, *C. arboreus*, *P. tokejii*, and *P. rhanis* (Ueda et al. 2009; Okada and Suda 2012; Sato et al. 2018; Takakuwa et al. 2007). However, *C. procelurus*, which is the largest carabid species that was captured in Ooyamazawa, was not negatively affected, even though previous studies have reported that *C. procelurus* is highly sensitive to

Fig. 10.7 Effect of mean body mass on changes in the annual catch of dominant carabid species. Vertical values were calculated using the coefficients of Poisson regression for each species (Fig. 10.6)

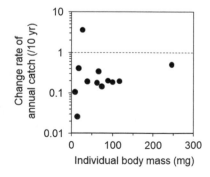

Fig. 10.8 Annual variation in the species composition of carabids captured in the Ooyamazawa riparian forest. Data were ordinated using principal component analysis (PCA), with partialing out the effect of subplot, and rare species (<10 individuals) were excluded from analysis. (**a**) Mean and standard error of site scores among subplots in each year. (**b**) Species score for each carabid species. Species names are abbreviated by letters: NbSa, *Nebria sadona*; CaAl, *Carabus albrechti*; CaPr, *Carabus procerulus*; CaAr, *Carabus arboreus*; CaHa, *Carabus harmandi*; TrYo, *Trichotichnus yoshiroi*; NgBe, *Negreum bentonis*; MyCu, *Myas cuprescens*; PtSu, *Pterostichus subovatus*; PtKt, *Pterostichus katashinensis*; PtMi, *Pterostichus mitoyamanus*; PtOk, *Pterostichus okutamae*; PtRh, *Pterostichus rhanis*; PtTo, *Pterostichus tokejii*; SyAg, *Synuchus agonus*; SyMe, *Synuchus melantho*

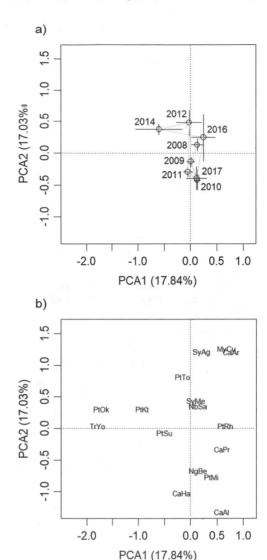

reductions in forest floor vegetation cover (Sato et al. 2018). It is possible that the *C. procelurus* population had already been reduced by vegetation reductions that occurred before the monitoring project was initiated and that the resulting low density was maintained thereafter or that the effect of increased temperature mitigated the negative impact of deer browsing (discussed below).

The last decade was the hottest one among the last 90 years in the Chichibu region. Although the mean annual air temperature of the region has exhibited constant fluctuation ($\pm 1°C$, until the late 1980s), it subsequently increased by ~1°C during the next 10 years and has remained high since 2000, as documented by the weather station in Chichibu (Fig. 10.1). Such long-lasting warm conditions in these years may affect the carabid community. However, if warming was the main influencing factor of the observed community changes, the observed trends would be expected to differ between species with different temperature preferences. In general, species that are distributed at higher latitudes or altitudes (i.e., adapted to colder climates) are more negatively affected by warming than other coexisting species that are adapted to warmer climates. According to Ishikawa (1986) and Sasaki and Chishima (1991), who investigated the vertical distribution (150–2100 m alt.) of *Carabus* spp. in the Kanto Mountains, *C. arboreus ogurai* and *C. harmandi okutamaensis* are distributed at higher altitudes (850–2100 m), whereas *C. albrechti esakianus* and *C. procelurus procelurus* are distributed at lower altitudes (300–2000 m). In addition, only the two lower-altitude species were found in the lowland forests at the foot of the Kanto Mountains (90–230 m; Matsumoto 2005, 2009, 2012; Soga et al. 2013), and only *C. p. procelurus* was found in the forests of the Kanto Plain at lower altitude (20–90 m) and similar latitudes (Shibuya et al. 2008, 2011, 2014; Taniwaki et al. 2005a). In Ooyamazawa, *C. a. ogurai*, *C. h. okutamaensis*, and *C. a. esakianus* exhibited similar decreasing trends, whereas *C. p. procelurus* did not (Fig. 10.6), and another species that was found in lowland forest and was also dominant in Ooyamazawa was *Myas cuprescens* (Matsumoto 2005; Shibuya et al. 2008, 2011, 2014; Soga et al. 2013; Taniwaki et al. 2005b), which also failed to exhibit any significant change (Fig. 10.6). These two species (*C. p. procelurus* and *M. cuprescens*), which can survive in warm lowland forests, might be positively affected by temperature increases, thereby compensating for the negative effects of deer overabundance. However, distinct directional changes in species composition were not detected by the principal component analysis (Fig. 10.8). It is possible that the warming effects were overwhelmed by the overall negative impacts of deer overabundance or that the species composition might have already reached a new equilibrium after the major warming event that occurred from the late 1980s to the 1990s, before the initiation of beetle monitoring in Ooyamazawa. The species that are dominant in lowland forests and are also reported to respond positively to deer-mediated reductions in forest floor vegetation (e.g., *P. yoritomus* and *S. cycloderus*) are currently rare in Ooyamazawa but could increase with additional increases in temperature.

Temperature changes can affect carabid body size. For example, certain *Carabus* spp. are reportedly larger in warmer regions or at lower altitudes (Sota et al. 2000a, b). This is likely owing to adaptation or plastic responses to climatic conditions or prey availability. Among the dominant carabid species in Ooyamazawa, *C. procelurus*, *P. tokejii*, and *S. melantho* exhibited increasing trends in individual body mass (Fig. 10.9), and the growth rate of *C. procelurus*, in particular, which is distributed in warm lowland forests, might have been enhanced by the temperature increase. Meanwhile, it is also possible that body size increases were caused by the relaxation of inter- or intraspecific competition. In contrast, the body size of *P. subovatus* was observed to decrease, which may reflect an intensification of intraspecific competition, owing to slight increases in the species' abundance (Figs. 10.6 and 10.9).

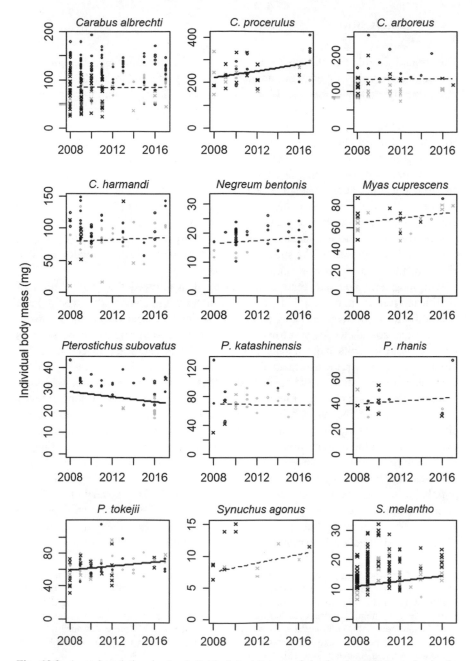

Fig. 10.9 Annual variation in the individual body mass of dominant carabid species in the Ooyamazawa riparian forest. Damaged individuals were excluded. Circle, Spring–Summer (May–June); cross, Autumn (September–October); grey, male; black, female. Solid and dashed lines indicate significant and insignificant regressions by year, respectively (likelihood ratio test, $\alpha = 0.05$). "Sampling season" and "sex" were used as random factors

10.5   Conclusion

The ground-dwelling beetle survey revealed that the Ooyamazawa riparian forest harbors a remarkably diverse and unique community of carabid beetles, including many locally endemic species. The characteristics of this community could be supported by the well-preserved stable forest environment, diverse microenvironments within riparian areas, huge species pool of forest-inhabiting carabids in central Honshu, and the geohistory of the Japanese Archipelago.

Sadly, the 10-year monitoring project also confirmed drastic or critical reductions in most species that prevailed throughout the plot. Two profound and long-lasting environmental pressures that have been introduced during the last two to three decades (i.e., deer-mediated reductions in forest floor vegetation and temperature increase) may have interactively caused the observed declines. If the present situation persists, most species, including many local endemics, will be extirpated from the site and be replaced by the species that are dominant at lower altitudes and that either tolerate or prefer deer-related disturbances.

The long-term and quantitative monitoring data allowed the identification of significant trends in the carabid beetle community that, generally, would have been difficult to understand using short-term or less frequent observation, mostly because the abundance of short-lived and rapidly reproducing animals, such as insects, often exhibits large annual fluctuations. It is necessary to continue this monitoring, in order to document the response of the beetle community to the persistence or alteration of the current deer density and climate conditions. Moreover, the observation of such pronounced species declines demonstrates the urgency of conserving the Ooyamazawa ecosystem. Indeed, Sato et al. (2018) described the effectiveness of providing early protection for forest floor communities against deer overbrowsing in the preservation of carabid species that are sensitive to declines in vegetation. A conservation project, the goal of which is to install deer exclosures in this area, has been just launched by the prefectural office, and the exclusion and population management of deer are expected to promote the recovery of the forest floor vegetation and, subsequently, of carabids and other vegetation-dependent animals. The monitoring data presented here also provides a basis for estimating the extent and direction of carabid community recovery.

Acknowledgments I am grateful to Ayu Toyota, for the launch and management of the beetle census in the first year of the monitoring, to Kôji Sasakawa and Toshio Kishimoto, for beetle identification, to Hitomi Asano and Keiko Ono, for laboratory work, and to Takafumi Hino and Hiroshi Ikeda, for helpful advice regarding statistical analyses and the manuscript. I am also grateful to all the individuals that contributed to the fieldwork: Hitoshi Sakio, Motohiro Kawanishi, Motoki Higa, Masako Kubo, Nobuhiko Wakamatsu, Kazuhiko Kato, Susumu Yamanaka, Nobuki Eikoshi, Ryo Kitagawa, and Atsuko Fukamachi.

References

Allombert S, Stockton S, Martin JL (2005) A natural experiment on the impact of overabundant deer on forest invertebrates. Conserv Biol 19:1917–1929

Bardgett RD, Wardle DA (2003) Herbivore-mediated linkages between aboveground and belowground communities. Ecology 84:2258–2268

Bressette JW, Beck H, Beauchamp VB (2012) Beyond the browse line: complex cascade effects mediated by white-tailed deer. Oikos 121:1749–1760

Brooks DR, Bater JE, Clark SJ, Monteith DT, Andrews C, Corbett SJ, Beaumont DA, Chapman JW (2012) Large carabid beetle declines in a United Kingdom monitoring network increases evidence for a widespread loss in insect biodiversity. J Appl Ecol 49:1009–1019

Côté SD, Rooney TP, Tremblay JP, Dussault C, Waller DM (2004) Ecological impacts of deer overabundance. Annu Rev Ecol Evol Syst 35:113–147

Dobson M, Kawamura Y (1998) Origin of the Japanese land mammal fauna: allocation of extant species to historically-based categories. Quartern Res 37:385–395

Flowerdew JR, Ellwood SA (2001) Impacts of woodland deer on small mammal ecology. Forestry 74:277–287

Foster CN, Barton PS, Lindenmayer DB (2014) Effects of large native herbivores on other animals. J Appl Ecol 51:929–938

Habu A (1958) Study on the species of the subgenus *Rhagadus* of *Pterostichus* from Japan. Mushi 31:1–13

Habu A (1978) Fauna Japonica Carabidae: Platynini (Insecta: Coleoptera). Keigaku Publishing Co., Tokyo

Hoekman D, Levan KE, Ball GE, Browne RA, Davidson RL, Erwin TL, Knisley CB, Labonte JR, Lundgren J, Maddison DR, Moore W, Niemelä J, Ober KA, Pearson DL, Spence JR, Will K, Work T (2017) Design for ground beetle abundance and diversity sampling within the National Ecological Observatory Network. Ecosphere 8:e01744

Imura Y, Mizusawa K (2013) The Carabus of Japan. Roppon-Ashi Entomological Books, Tokyo

Ishihara MI, Suzuki SN, Nakamura M, Enoki T, Fujiwara A, Hiura T, Homma K, Hoshino D, Hoshizaki K, Ida H, Ishida K, Itoh A, Kaneko T, Kubota K, Kuraji K, Kuramoto S, Makita A, Masaki T, Namikawa K, Niiyama K, Noguchi M, Nomiya H, Ohkubo T, Saito S, Sakai T, Sakimoto M, Sakio H, Shibano H, Sugita H, Suzuki M, Takashima A, Tanaka N, Tashiro N, Tokuchi N, Yoshida T, Yoshida Y (2011) Forest stand structure composition and dynamics in 34 sites over Japan. Ecol Res 26:1007–1008

Ishikawa R (1986) Distribution of ground beetles (Coleoptera: Carabidae: Carabinae: Carabina) in the upper and middle regions of Tama River Basin. In: Ishikawa R (ed) Analysis of distribution patterns of low-mobile animals in Tama River Basin. The Tokyu Foundation for Better Environment, Tokyo, pp 3–19. (in Japanese)

Kawamura Y (2007) Last glacial and Holocene land mammals of the Japanese Islands: their fauna, extinction and immigration. Quatern Res 46:171–177

Koivula M (2011) Useful model organisms, indicators, or both? Ground beetles (Coleoptera: Carabidae) reflecting environmental conditions. ZooKeys 100:287–317

Lessard JP, Reynolds WN, Bunn WA, Genung MA, Cregger MA, Felker-Quinn E, Barrios-Garcia MN, Stevenson ML, Lawton RM, Brown CB, Patrick M, Rock JH, Jenkins MA, Bailey JK, Schweitzer JA (2012) Equivalence in the strength of deer herbivory on above and below ground communities. Basic Appl Ecol 13:59–66

Löbl I, Löbl D (2017) Catalogue of Palaearctic Coleoptera Vol. 1: Archostemata - Myxophaga - Adephaga, revised and updated edition. Brill, Leiden, Netherlands

Lövei L, Sunderland KD (1996) Ecology and behavior of ground beetles (Coleoptera: Carabidae). Annu Rev Entomol 41:231–256

Matsumoto K (2005) Ground beetle assemblages and the management of forest understory in the Tama Experimental Station of the Forestry and Forest Products Research Institute and Tokyo Metropolitan Sakuragaoka Park. Jpn J Environ Entomol Zool 16:31–38. (in Japanese with English summary)

Matsumoto K (2009) Ground beetle assemblages in the forests of the Field Museum Tama Hills of the Tokyo University of Agriculture and Technology and Tokyo Metropolitan Nanao Park with particular attention to the effect of clearing of understory. Jpn J Environ Entomol Zool 20:115–125. (in Japanese with English summary)

Matsumoto K (2012) Ground beetle assemblages in the forests of Tobuki-kita conservation area with special reference to effects of the understory management. Jpn J Environ Entomol Zool 23:9–17. (in Japanese with English summary)

Melis C, Buset A, Aarrestad PA, Hanssen O, Meisingset EL, Andersen R, Moksnes A, Røskaft E (2006) Impact of red deer *Cervus elaphus* grazing on bilberry *Vaccinium myrtillus* and composition of ground beetle (Coleoptera, Carabidae) assemblage. Biodivers Conserv 15:2049–2059

Melis C, Sundby M, Andersen R, Moksnes A, Pedersen B, Røskaft E (2007) The role of moose *Alces alces* L. in boreal forest the effect on ground beetles (Coleoptera, Carabidae) abundance and diversity. Biodivers Conserv 16:1321–1335

Morecroft MD, Bealey CE, Beaumont DA, Benham S, Brooks DR, Burt TP, Critchley CNR, Dick J, Littlewood NA, Monteith DT, Scott WA, Smith RI, Walmsey C, Watson H (2009) The UK environmental change network: emerging trends in the composition of plant and animal communities and the physical environment. Biol Conserv 142:2814–2832

Morita S (1997) The group of *Trichotichnus leptopus* (Coleoptera, Carabidae) of Japan. Elytra 25:521–585

Niwa S, Toyota A, Kishimoto T, Sasakawa K, Abe S, Chishima T, Higa M, Hiura T, Homma K, Hoshino D, Ida H, Kamata N, Kaneko Y, Kawanishi M, Kobayashi K, Kubota K, Kuraji K, Masaki T, Niiyama K, Noguchi M, Nomiya H, Saito S, Sakimoto M, Sakio H, Sato S, Shibata M, Takashima A, Tanaka H, Tashiro N, Tokuchi N, Torikai H, Yoshida T (2016) Monitoring of the ground-dwelling beetle community and forest floor environment in 22 temperate forests across Japan. Ecol Res 31:607–608

Okada T, Suda K (2012) Effect of different forest floor environment on the carabid (Coleoptera: Carabidae) community structure in Oku-Nikko, Tochigi Prefecture, central Japan. Bull Geo Environ Sci 14:1–6

Okuzaki Y, Tayasu I, Okuda N, Sota T (2009) Vertical heterogeneity of a forest floor invertebrate food web as indicated by stable isotope analysis. Ecol Res 24:1351–1359

Pozsgai G, Littlewood NA (2014) Ground beetle (Coleoptera: Carabidae) population declines and phenological changes: is there a connection? Ecol Indic 41:15–24

Rainio J, Niemelä J (2003) Ground beetles (Coleoptera: Carabidae) as bioindicators. Biodivers Conserv 12:487–506

Rooney TP, Waller ÐM (2003) Direct and indirect effects of white-tailed deer in forest ecosystems. For Ecol Manag 181:165–176

Sakio H, Kubo M, Kawanishi M, Higa M (2013) Effects of deer feeding on forest floor vegetation in the Chichibu Mountains, Japan. J Jpn Soc Reveget Tech 39:226–231. (in Japanese with English summary)

Sasakawa K (2005a) Phylogenetic studies of the subgenus *Nialoe* (s. lat.) (Coleoptera, Carabidae, genus *Pterostichus*), part 2: the *Asymmetricus* species group. Zool Sci 22:1217–1228

Sasakawa K (2005b) *Pterostichus macrogenys* Bates, 1883 (Coleoptera, Carabidae) and its allied species of northern Japan. Biogeography 7:69–78

Sasakawa K (2009) *Pterostichus macrogenys* Bates (Coleoptera: Carabidae) and its allied species of northern Japan: descriptions of seven additional species and possible evidence supporting species status. Zool Stud 48:262–269

Sasakawa K (2016) Two new species of the ground beetle subgenus *Sadonebria* Ledoux & Roux, 2005 (Coleoptera, Carabidae, *Nebria*) from Japan and first description of larvae of the subgenus. ZooKeys 578:97–113

Sasakawa K, Kubota K (2009) Phylogeny of ground beetles subgenus *Nialoe* (s. lat.) Tanaka (Coleoptera: Carabidae; genus *Pterostichus*): a molecular phylogenetic approach. Entomol Sci 12:308–313

Sasaki K, Chishima T (1991) Vertical distribution and habitat selection of carabid beetles in upper Chichibu. Misc Inf Univ Tokyo For 28:1–12. (in Japanese)

Sato S, Suzuki M, Taniwaki T, Tamura A (2018) Indirect effects of increasing sika deer (*Cervus nippon*) on carabid beetles in the Tanzawa mountains. J Jpn For Soc 100:141–148. (in Japanese with English summary)

Shibuya S, Kubota K, Kikvidze Z, Ohsawa M (2008) Differential sensitivity of ground beetles, *Eusilpha japonica* and Carabidae, to vegetation disturbance in an abandoned coppice forest in central Japan. Euras J For Res 11:61–72

Shibuya S, Kubota K, Ohsawa M, Kikvidze Z (2011) Assembly rules for ground beetle communities: what determines community structure, environmental factors or competition? Eur J Entomol 108:453–459

Shibuya S, Kikvidze Z, Toki W, Kanazawa Y, Suizu T, Yajima T, Fujimori T, Mansournia MR, Sule Z, Kubota K, Fukuda K (2014) Ground beetle community in suburban Satoyama — a case study on wing type and body size under small scale management. J Asia-Pac Entomol 17:775–780

Shibuya S, Kiritani K, Fukuda K (2018) Hind wings in ground beetles (Coleoptera: Carabidae and Brachinidae) – morphology, length, and characteristics of each subfamily. J Jpn Soc Ecol 68:19–41. (in Japanese with English summary)

Soga M, Kanno N, Yamaura Y, Koike S (2013) Patch size determines the strength of edge effects on carabid beetle assemblages in urban remnant forests. J Insect Conserv 17:421–428

Sota T, Nagata N (2008) Diversification in a fluctuating island setting: rapid radiation of *Ohomopterus* ground beetles in the Japanese Islands. Phil Trans R Soc B 363:3377–3390

Sota T, Takami Y, Kubota K, Ujiie M, Ishikawa R (2000a) Interspecific body size differentiation in species assemblages of the carabid beetles *Ohomopterus* in Japan. Popul Ecol 42:279–291

Sota T, Takami Y, Kubota K, Ishikawa R (2000b) Geographic variation in the body size of some Japanese *Leptocarabus* species (Coleoptera, Carabidae): the "toppled-domino pattern" in species along a geographic cline. Entomol Sci 3:309–320

Stewart AJA (2001) The impact of deer on lowland woodland invertebrates: a review of the evidence and priorities for future research. Forestry 74:259–270

Suominen O (1999) Impact of cervid browsing and grazing on the terrestrial gastropod fauna in the boreal forests of Fennoscandia. Ecography 22:651–658

Suominen O, Danell K (1999) Indirect effects of mammalian browsers on vegetation and ground-dwelling insects in an Alaskan floodplain. Écoscience 6:505–510

Suominen O, Niemelä J, Martikainen P, Niemelä P, Kojola I (2003) Impact of reindeer grazing on ground-dwelling Carabidae and Curculionidae assemblages in Lapland. Ecography 26:503–513

Suzuki SN, Ishihara MI, Nakamura M, Abe S, Hiura T, Homma K, Higa M, Hoshino D, Hoshizaki K, Ida H, Ishida K, Kawanishi M, Kobayashi K, Kuraji K, Kuramoto S, Masaki T, Niiyama K, Noguchi M, Nomiya H, Saito S, Sakai T, Sakimoto M, Sakio H, Sato T, Shibano H, Shibata M, Suzuki M, Takashima A, Tanaka H, Takagi M, Tashiro N, Tokuchi N, Yoshida T, Yoshida Y (2012) Nation-wide litter fall data from 21 forests of the monitoring sites 1000 project in Japan. Ecol Res 27:989–990

Tada R (1994) Paleoceanographic evolution of the Japan sea. Palaeogeogr Palaeoclimatol Palaeoecol 108:487–508

Takakuwa M, Fukada S, Fujita H (2007) Insect fauna and its density inside and outside the deer-fence at Mitsumine of Mt. Tanzawa, Central Japan. In: The research group of the Tanzawa-Ohyama Mountains (ed) Scientific reports of comprehensive research on the Tanzawa-Ohyama mountains. Hiraoka Environmental Science Laboratory, Sagamihara, pp 227—231. (in Japanese)

Takatsuki S (2009) Effects of sika deer on vegetation in Japan: a review. Biol Conserv 142:1922–1929

Taniwaki T, Kuno H, Hosoda H (2005a) Annual changes of community structure of ground insects at the isolated small stands in suburbs of Tokyo. J Jpn Soc Reveget Tech 30:552–560. (in Japanese with English summary)

Taniwaki T, Kuno H, Kishi Y (2005b) Comparison of ground beetle fauna on managed and short and long-term unmanaged floors in a suburban forest. J Jpn Soc Reveget Tech 31:260–268. (in Japanese with English summary)

Terui A, Akasaka T, Negishi JN, Uemura F, Nakamura F (2017) Species-specific use of allochthonous resources by ground beetles (Carabidae) at a river–land interface. Ecol Res 32:27–35

Thiele HU (1977) Carabid beetles in their environments: a study on habitat selection by adaptations in physiology and behaviour. Springer, Berlin

Tominaga O, Su Z-H, Kim C-G, Okamoto M, Imura Y, Osawa S (2000) Formation of the Japanese Carabina fauna inferred from a phylogenetic tree of mitochondrial ND5 gene sequences (Coleoptera, Carabidae). J Mol Evol 50:541–549

Ueda A, Hino T, Ito H (2009) Relationships between browsing on dwarf bamboo (*Sasa nipponica*) by sika deer and the structure of ground beetle (Coleoptera: Carabidae) assemblage. J Jpn For Res 91:111–119. (in Japanese with English summary)

Ueno S, Kurosawa Y, Satoh M (1985) The Coleoptera of Japan in color, vol 2. Hoikusha, Osaka. (in Japanese)

Vanbergen AJ, Woodcock BA, Koivula M, Niemelä J, Kotze DJ, Bolger T, Golden V, Dubs F, Boulanger G, Serrano J, Lencina JL, Serrano A, Aguiar C, Grandchamp A-C, Stofer S, Szél G, Ivits E, Adler P, Markus J, Watt AD (2010) Trophic level modulates carabid beetle responses to habitat and landscape structure: a pan-European study. Ecol Entomol 35:226–235

Wardle DA, Barker GM, Yeates GW, Bonner KI, Ghani A (2001) Introduced browsing mammals in New Zealand natural forests: aboveground and belowground consequences. Ecol Monogr 71:587–614

Yamada H, Takatsuki S (2015) Effects of deer grazing on vegetation and ground-dwelling insects in a larch forest in Okutama, Western Tokyo. Int J For Res, Article ID 687506

Zhang A-B, Sota T (2007) Nuclear gene sequences resolve species phylogeny and mitochondrial introgression in *Leptocarabus* beetles showing trans-species polymorphisms. Mol Phylogenet Evol 45:534–546

Chapter 11
Avifauna at Ooyamazawa: Decline of Birds that Forage in Bushy Understories

Mutsuyuki Ueta

Abstract The breeding and wintering bird fauna at Ooyamazawa were monitored by point counting every year from 2010 to 2017. The abundance of bird species that feed in bushes, including the Japanese Bush Warbler (*Cettia diphone*) and Siberian Blue Robin (*Luscinia cyane*), decreased. On the contrary, the other bird species that feed in habitats other than bushy understory did not decrease significantly. In Ooyamazawa, deers have serious impact on the understory vegetation. Thus, deer browsing likely affected the number of birds using the understory vegetation.

Keywords Bird fauna · Climate · Deer grazing · Stream noise · Understory vegetation

11.1 Introduction

The "Monitoring Sites 1000 Project" was launched by the Japanese Ministry of the Environment to monitor the natural environment in Japan. Birds are an important monitoring target taxon in the project because they are easy to survey, and a database on their past populations is available. The bird database provides important information on the changes in abundance, species composition, and relative distributions in the birds' habitat. Therefore, starting in 2010, I started a bird survey project in Ooyamazawa.

In this section, I will describe the features of the avifauna at Ooyamazawa's riparian forest and the impact of deer browsing on the avifauna during eight years of monitoring.

M. Ueta (✉)
The Japan Bird Research Association, Tokyo, Japan
e-mail: mj-ueta@bird-research.jp

© The Author(s) 2020
H. Sakio (ed.), *Long-Term Ecosystem Changes in Riparian Forests*, Ecological Research Monographs, https://doi.org/10.1007/978-981-15-3009-8_11

11.2   Characteristics of Avifauna at Ooyamazawa

The bird survey in Ooyamazawa was conducted from 2010 to 2017. I placed five fixed survey points within the study site, and the survey points were visited twice every breeding and wintering season. During the 8 years, I recorded a total of 44 and 23 species during the breeding season and wintering season, respectively (Table 11.1, Fig. 11.1).

11.2.1   Effects of Topography and Climate on Bird Fauna

The survey points in Ooyamazawa were located in the north-facing slope of a valley. The environment was cold and harsh, especially in winter. This may have been a factor affecting the avifauna of this study site.

I compared the avifauna of Ooyamazawa with that of the University of Tokyo Chichibu Forest (University Forest) situated on a southern slope about 5 km away from Ooyamazawa to compare the environmental effect. The University Forest is located at an altitude of about 1200 m and is covered by a deciduous broad-leaved forest dominated by *Fagus japonica* and *F. crenata* mixed with *Tsuga sieboldii*. The forest floor had once been dominated by *Sasa borealis* before deer browsing intensified. Currently, the forest floor is mostly denuded except for sporadic patches of *Pieris japonica*, which deers do not eat.

In the 2016 and 2017 breeding seasons, the total number of birds recorded in Ooyamazawa were 53 and 52, respectively, whereas those of the University Forest were 48 and 48, respectively. There were no differences between the abundances in the two study sites during the breeding season. However, during the wintering season of 2016 and 2017, the bird abundances in Ooyamazawa were 30 and 40, respectively, while those in the University Forest were 109 and 71, respectively. Overall, the birds in Ooyamazawa tended to decrease in the winter, while those in the University Forest tended to increase in the winter (Fig. 11.2).

I compiled the bird data based on the birds' feeding habitat niche, to see the effect of the environment on their foraging behavior. During the wintering season at Ooyamazawa, ground-foraging birds decreased, while stem-foraging birds remained stable, and canopy-foraging birds decreased slightly (Fig. 11.2). At the same time, in the University Forest, all birds increased, but especially the ground-foraging birds, which increased greatly (Fig. 11.2).

The study site at Ooyamazawa is covered with about 30 cm of snow in winter because of its location on the northern slope of a valley. On the other hand, the University Forest is located on the southern slope, which has little snow cover except during snowfall events.

In Ooyamazawa, the decrease in the abundance of ground-feeding birds in winter may be explained by the difficulty of feeding on snow-covered ground. On the other hand, because the University Forest is relatively free of snow, the ground-feeding

Table 11.1 List of birds recorded at Ooyamazawa

English name	Scientific name	Summer	Winter	Forage	Nest
Copper Pheasant	*Syrmaticus soemmerringii*	△	△	G	B
Japanese Green Pigeon	*Treron sieboldii*	◎		T	T
Rufous Hawk-Cuckoo	*Hierococcyx hyperythrus*	○		T	P
Lesser Cuckoo	*Cuculus poliocephalus*	○		T	P
Oriental Cuckoo	*Cuculus optatus*	◎		T	P
White-throated Needletailed Swift	*Hirundapus caudacutus*	R		A	-
Pacific Swift	*Apus pacificus*	R		A	-
Eurasian Sparrowhawk	*Accipiter nisus*	△		A	T
Oriental Scops Owl	*Otus sunia*	△		A	C
Japanese Pygmy Woodpecker	*Dendrocopos kizuki*	◎	○	S	C
White-backed Woodpecker	*Dendrocopos leucotos*	○	○	S	C
Great Spotted Woodpecker	*Dendrocopos major*	○	△	S	C
Japanese Green Woodpecker	*Picus awokera*	○	△	S	C
Eurasian Jay	*Garrulus glandarius*	◎	○	T	T
Large-billed Crow	*Corvus macrorhynchos*	○	○	G	T
Willow Tit	*Poecile montanus*	○	◎	T	C
Varied Tit	*Poecile varius*	○	○	T	C
Coal Tit	*Periparus ater*	◎	◎	T	C
Japanese Tit	*Parus minor*	◎	△	T	C
Brown-eared Bulbul	*Hypsipetes amaurotis*	△	△	T	T
Japanese Bush Warbler	*Cettia diphone*	◎		B	B
Asian Stubtail	*Urosphena squameiceps*	△		B	B
Long-tailed Tit	*Aegithalos caudatus*	○	○	T	T
Japanese Leaf Warbler	*Phylloscopus xanthodryas*	R		T	–
Sakhalin Leaf Warbler	*Phylloscopus borealoides*	◎		T	B
Eastern Crowned Leaf Warbler	*Phylloscopus coronatus*	◎		T	B
Japanese White-eye	*Zosterops japonicus*	△		T	T
Eurasian Nuthatch	*Sitta europaea*	◎	◎	S	C
Eurasian Treecreeper	*Certhia familiaris*	○	○	S	C
Eurasian Wren	*Troglodytes troglodytes*	◎	○	G	G
Brown Dipper	*Cinclus pallasii*	△	△	G	G
Siberian Thrush	*Zoothera sibirica*	○		G	T
Scaly Thrush	*Zoothera dauma*	△		G	T
Japanese Thrush	*Turdus cardis*	△		G	T
Brown-headed Thrush	*Turdus chrysolaus*	△		G	T
Naumann's Thrush	*Turdus naumanni*		○	T	–
Japanese Robin	*Luscinia akahige*	◎		B	B
Siberian Blue Robin	*Luscinia cyane*	◎		B	B

(continued)

Table 11.1 (continued)

English name	Scientific name	Summer	Winter	Forage	Nest
Red-flanked Bluetail	*Tarsiger cyanurus*	R		G	–
Narcissus Flycatcher	*Ficedula narcissina*	○		T	C
Blue-and-white Flycatcher	*Cyanoptila cyanomelana*	◎		A	G
Grey Wagtail	*Motacilla cinerea*	△		G	G
Brambling	*Fringilla montifringilla*		△	T	–
Eurasian Siskin	*Carduelis spinus*		△	T	–
Asian Rosy Finch	*Leucosticte arctoa*		△	G	–
Eurasian Bullfinch	*Pyrrhula pyrrhula*		◎	T	–
Japanese Grosbeak	*Eophona personata*	△	○	G	T
Grey Bunting	*Emberiza variabilis*	R		G	–
Red-billed Leiothrix	*Leiothrix lutea*	◎		B	B

◎ dominant, ○ common, △ few, R rare
Forage: *G* ground-foraging, *B* bush, *S* stem, *T* tree, *A* air
Nest: *G* ground cliff, *B* bush, *C* cavity, *T* tree

birds over-wintering in this site may find it relatively easier to forage on the ground. The reason that the abundance of tree stem-feeding birds in Ooyamazawa did not change through the breeding and wintering seasons may be because tree stems are hardly affected by snow cover and thus continues to provide a foraging microhabitat throughout the year.

11.2.2   Effect of Stream Noise on Bird Fauna

Noise masking is a process that interferes with the use of acoustic signals that are critical to many bird species (Brumm and Slabbekoorn 2005; Francis et al. 2009). Therefore, birds living in areas exposed to anthropogenic noise may experience reduced reproductive success, which may ultimately lead to the exclusion of species from an otherwise suitable habitat (Slabbekoorn and Ripmeester 2008). Because stream noise is also thought to affect the avifauna, I studied the effects of stream noise on the local bird distribution during the breeding and wintering seasons between 2009 and 2012 in Ooyamazawa. In the breeding season, acoustic signals are especially important because they are used to establish and defend territory, and to provide cues for mating. Therefore, stream noise is expected to have a greater effect on the local bird distribution. I established three study sites, each of which included survey points with (50.3–57.9 dB (A)) and without stream noise (35.0–40.1 dB (A)). I then counted the abundance of birds of each species within a radius of 50 m. Each study site was similar in terms of vegetation type and density.

Fig. 11.1 Dominant bird species observed at Ooyamazawa. (1) *Cettia diphone* (photo by Kaoru Fujii), (2) *Phylloscopus borealoides* (photo by Satoshi Miki), (3) *Troglodytes troglodytes* (photo by Hiroshi Uchida), (4) *Luscinia cyane* (photo by Hiroshi Uchida), (5) *Cyanoptila cyanomelana* (photo by Yoshihiko Yuasa), (6) *Leiothrix lutea* (photo by Yukitoshi Otsuka)

Fig. 11.2 Comparison of bird abundance in each site between seasons. The birds are classified into the following foraging niche groups: ground, stem, and canopy-foraging groups. Total numbers of observed birds during 2016 and 2017 are shown

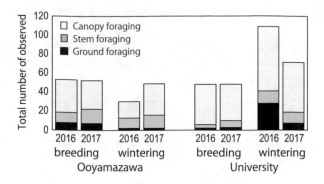

The sound levels (sound pressure waves) at each survey point were characterized by their frequency and decibel range on an unweighted decibel scale. Sonograms for each bird species encountered were created from data obtained during the study.

During the breeding season when song communication is important, Sakhalin Leaf Warblers (*Phylloscopus borealoides*) were significantly more abundant in the sites with stream noise, but Eurasian Nuthatches (*Sitta europaea*) and Coal Tits (*Periparus ater*) were significantly more abundant in the sites free from stream noise (Fig. 11.3). The numbers of Eurasian Wrens (*Troglodytes troglodytes*) and Blue-and-White Flycatchers (*Cyanoptitla cyanomelana*) did not differ significantly between the two types of sites. In the wintering seasons, however, the numbers of *S. europaea*, *P. ater* and Willow Tits (*Poecile montanus*) did not differ significantly between the sites with and without stream noise (Fig. 11.3).

Phylloscopus borealoides and *T. troglodytes* sang at frequencies higher than 6000 Hz (Fig. 11.4); thus, they were little affected by the stream noise, because the sound pressure of stream noise was reduced at sound frequencies above 6000 kHz (Ueta 2012). On the other hand, *S. europaea* and *P. ater* sang at frequencies lower than 6000 Hz. Therefore, the songs of these two species were greatly affected by stream noise.

Except for the stream noise, the general features of the environment in each study site were similar. It is likely that bird species that sing in the frequency range overlapping the loudest stream noise moved away from the noisy stream during the breeding season when song information is critically important for them. Birds that sang in a higher frequency seemed either to prefer the stream-side locations or to remain on the stream-side sites where the competition is low.

Although it is possible that bird sounds were unrecognizable due to the loud stream noise, the range of recorded sounds was only within 50 m; at this range, it has been confirmed that the observer can easily hear bird songs even in the presence of loud stream noise. Therefore, the effect of stream noise is more likely the result of a behavioral shift in birds rather than error on the part of the observers.

Many aquatic insects emerge along the river (Murakami 2001), which are an important food source for birds in the season when trees are without leaves. In Ooyamazawa where the bud-break occurs late, the tree leaves were not open during

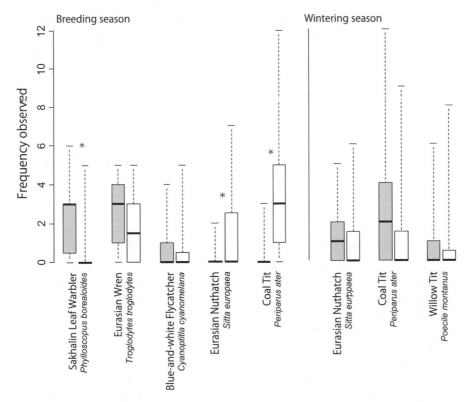

Fig. 11.3 Comparison in observed occurrences of bird species between sites with (shaded bar) and without (open bar) stream noise. *GLMM $P < 0.05$

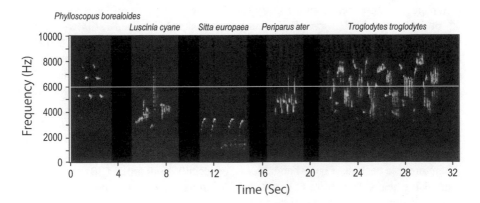

Fig. 11.4 Sonogram of the songs of dominant bird species

the study's breeding period. Therefore, the abundance of leaf-feeders such as Lepidoptera larvae was likely low, which adds to the importance of emerging aquatic insects from the stream as the major food source for birds that forage for flying insects. If the availability of food for specialized insectivorous birds is more favorable on the stream-side, then it is possible that *P. borealoides* occurs more frequently in sites with stream noise even at the cost of the stream noise interfering with bird communication. However, for other bird species that do not benefit from flying insects, there is a cost to living in the stream-side, and thus these species are less abundant.

These results suggest that stream noise affects the avifauna of Ooyamazawa in the breeding season by hindering communication through song.

11.3   Decrease of Bird Species that Feed in Bushes

In Ooyamazawa, the forest understory, especially *Sasa borealis*, has decreased due to the impact of browsing by Sika Deer (*Cervus nippon*) since 2000 (Sakio et al. 2013, Chap. 8). To investigate the impact of browsing by *C. nippon* on avifauna, I analyzed the population trends of the 15 most common bird species that were recorded in more than 20 surveys. Population indices and long-term trends for each species were calculated using TRIM (Trends and Indices for Monitoring data, Package "rtrim 2.0.4" Bogaart et al. 2018).

The abundance of bird species that feed in bushes, namely the Japanese Bush Warbler (*Cettia diphone*), Siberian Blue Robin (*Luscinia cyane*), Japanese Robin (*Luscinia akahige*), and Red-billed Leiothrix (*Leiothrix lutea*) decreased significantly during the 8 years of monitoring (Fig. 11.5). By 2017, these birds had almost completely disappeared from the study site. The abundance of bird species that nest in bushes, namely *P. borealoides* and the Eastern Crowned Willow Warbler (*Phylloscopus coronatus*) also decreased, but they were still observed in 2017.

On the contrary, the other bird species that feed in habitats other than bushy understory did not decrease significantly (Fig. 11.6). The levels of the White-bellied Green pigeon (*Treron sieboldii*), *T. troglodytes*, and *C. cyanomelana* increased significantly.

After the loss of the Ooyamazawa forest understory, especially *S. borealis*, due to the impact of browsing by *C. nippon*, *S. borealis* experienced a bloom in 2013. After this, however, the remaining *S. borealis* died and have not revived since then. Therefore, the decline of bush-habitat birds, especially from around 2013, was likely caused by heavy deer browsing and the blooming-dieback phase of *S. borealis*. Similar declines in bird populations that feed in bushes have been observed in other forests in Japan that have experienced heavy deer browsing pressure (Hino 2006, Ueta et al. 2014).

Phylloscopus borealoides and *P. coronatus* that feed in the tree canopy and nest in bushes decreased but were still observed in 2017. These birds are less dependent

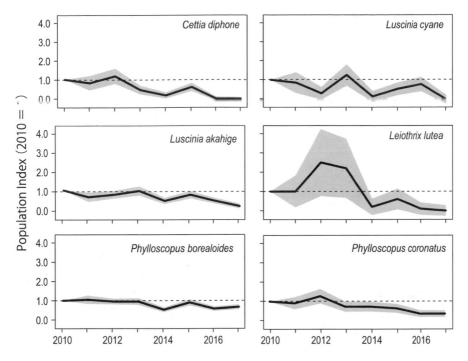

Fig. 11.5 The trends of population indices for six bird species that use bushes for feeding and/or nesting sites. Shaded areas show confidence intervals. The population index is presented with 2010 as the base year (=1)

on the forest understory compared to the bush-specialist species; thus, they have continued to inhabit the site.

Since the cessation of most forestry efforts in the 1970s, Japanese forests have matured. Therefore, at a nationwide scale, the abundance of bird species that depend on mature forests have also increased (Yamaura et al. 2009). Although summer visitor species experienced a rapid decline once in the 1980s and again in the 1990s (Higuchi and Morishita 1999), their populations are currently undergoing a restoration (Ueta 2016). This current general situation of Japanese forests may explain the significant increase in the populations of *T. sieboldii* and *C. cyanomelana*, as well as the stable population trend of other species.

The understory vegetation of the University Forest, located on a southern slope about five kilometers away from Ooyamazawa, was once also heavily affected by deer browsing. As a consequence, the former *S. borealis* understory is currently being replaced by *Pieris japonica* (Ericaceae), a species that deers do not eat. A similar change may occur in Ooyamazawa. The abundance levels of some of the birds using the forest understory will likely recover in response to these changes, although some species will not be able to respond. Since this is an ongoing process, it would be interesting to see what happens in the future through continued monitoring of the avifauna.

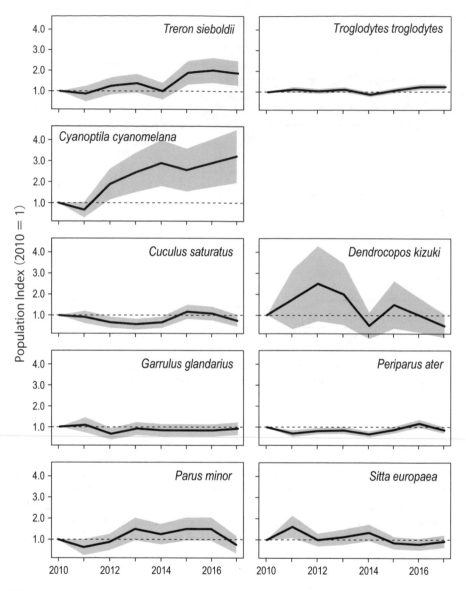

Fig. 11.6 The trends of population indices for nine bird species that do not use bushy understory for feeding and/or nesting sites. Shaded areas show confidence intervals. The population index is presented with 2010 as the base year (=1)

References

Bogaart P, van der Loo M, Pannekoek J (2018) rtrim: trends and Indices for Monitoring Data. https://cran.r-project.org/web/packages/rtrim/index.html

Brumm H, Slabbekoorn H (2005) Acoustic communication in noise. Adv Study Behav 35:151–209

Francis CD, Ortega CP, Cruz A (2009) Noise pollution changes avian communities and species interactions. Curr Biol 19:1415–1419

Higuchi H, Morishita E (1999) Population declines of tropical migratory birds in Japan. Actinia 12:51–59

Hino T (2006) The impact of herbivory by deer on forest bird community in Japan. Acta Zool Sin 52:684–686

Murakami M (2001) Bird community in riparian forest. Jpn J Ornithol 50:115–124

Sakio H, Kubo M, Kawanishi M, Higa M (2013) Effects of deer feeding on forest floor vegetation in the Chichibu Mountains, Japan. J Jpn Soc Reveget Tech 39:226–231

Slabbekoorn H, Ripmeester EA (2008) Birdsong and anthropogenic noise: implications and applications for conservation. Mol Ecol 17:72–83

Ueta M (2012) The effects of stream noise on local bird distribution during wintering and breeding seasons. Bird Res 8:S19–S24

Ueta M (2016) Result of "Breeding Bird Atlas" in 2016. Breed Bird Atlas Newsl 6:1–2

Ueta M, Iwamoto T, Nakamura Y, Kawasaki S, Konno S, Sato S, Takashi M, Takashima A, Takizawa K, Numano M, Harada O, Hirano T, Hotta M, Mikami K, Yanagida K, Matsui M, Arakida Y, Saiki M, Yukimoto S (2014) Monitoring the changes of land bird populations in Japan from 2009 to 2013. Bird Res 10:F1–F10

Yamaura Y, Amano T, Koizumi T, Mitsuda Y, Taki H, Okabe K (2009) Does land-use change affect biodiversity dynamics at a macroecological scale? A case study of birds over the past 20 years in Japan. Anim Conserv 12:110–119

Part V
Conclusion

Chapter 12
General Conclusion

Hitoshi Sakio

Abstract The research conducted at Ooyamazawa highlights the importance and value of long-term ecological research (LTER). The studies discussed here demonstrate that tree life-history strategies and regeneration are strongly correlated with disturbance regimes in riparian zones. Coexistence mechanisms among canopy tree species at Ooyamazawa reflect niche partitioning at early stages; however, long-term coexistence was more related to unpredictable large-scale disturbance events. LTER documented the unexpected effects of a rapid increase in sika deer populations. Deer browsing led to the decline of understory vegetation and trees at Ooyamazawa and the resulting cascade effects led to changes in the ground beetle and bird communities. Our 28-year research record indicated that global warming has impacted flowering and fruiting in *Fraxinus platypoda*. Our findings highlight the necessity of ongoing, long-term monitoring in capturing ecosystem change.

Keywords Avifauna · Cascade effect · Global warming · Ground beetle decline · Long-term ecological research · Ooyamazawa riparian forest · Phenology · Riparian forest conservation · Sika deer

12.1 Research at Ooyamazawa Riparian Forest Research Site

In the 1980s, long-term ecological research (LTER) sites were established across Japan. Research at the Ooyamazawa riparian forest research site was initiated early relative to other locations, in 1983 (Chap. 1); this included the establishment of the Ogawa Forest Reserve by the Forestry and Forest Products Research Institute (Nakashizuka and Matsumoto 2002). Pilot experiments were conducted from 1983 to 1986. In 1987, full-scale research began, focused on forest structure and

H. Sakio (✉)
Sado Island Center for Ecological Sustainability, Niigata University, Niigata, Japan
e-mail: sakio@agr.niigata-u.ac.jp; sakiohit@gmail.com

© The Author(s) 2020   215
H. Sakio (ed.), *Long-Term Ecosystem Changes in Riparian Forests*, Ecological
Research Monographs, https://doi.org/10.1007/978-981-15-3009-8_12

regeneration. In the same year, 20 seed traps were established within a core plot of 0.54 ha. Subsequent research regarding forest structure and seed production focused on the dominant canopy species *Fraxinus platypoda*. In 1991, the survey area was expanded to encompass 4.71 ha; research on canopy species was then broadened to include *Pterocarya rhoifolia* and *Cercidiphyllum japonicum*. In 1997, collaborating researchers began additional surveys. In December 2006, the site was registered within the Japan long-term ecological research network (JaLTER). In addition, a 1-ha plot, which included the existing core plot, was registered as a core site within the Ministry of Environment's Monitoring Sites 1000 Project.

Over 35 years, LTER at Ooyamazawa focused on the life-history strategies of riparian trees (Chap. 2) and coexistence mechanisms among riparian plants (Chaps. 6 and 7). Long-term phenological observations, which were made using seed traps and binoculars, revealed changes in the flowering and seed production cycles of *F. platypoda* (Chap. 2). During the course of the study period, dramatic changes occurred at Ooyamazawa. Sika deer (*Cervus nippon*) heavily influenced the understory vegetation after 2000, wherein overbrowsing caused a rapid decline in vegetation cover (Chap. 8). The effects of sika deer have been significant throughout Japan (Miura and Tokida 2008). The observed decline in understory vegetation at Ooyamazawa precipitated significant cascade effects on the ground beetle community (Chap. 10) and the local avifauna (Chap. 11).

12.2   Tree Life Histories and Dynamics

Based on the assumption that an understanding of life history is vital for understanding coexistence and ecosystem dynamics in forested stands, LTER at Ooyamazawa focused on tree life-history strategies within the riparian forest. These studies focused on the dominant canopy species *F. platypoda*, *P. rhoifolia*, and *C. japonicum*, as well as four *Acer* species (Chaps. 2–5). For each of the three dominant canopy species, the population structure and size and spatial distributions were quantified. Annual variation in seed production was also assessed, as were germination characteristics; notably, germination characteristics were investigated using both field and nursery observations. For the four *Acer* species, the sizes and spatial distributions of mature individuals, as well as characteristics of vegetative reproduction, were assessed.

Some tree species reproduce both sexually and asexually (i.e., vegetatively). Some species may favor vegetative reproduction in areas where their growth is limited, such as at the thresholds of their tolerance to light and substrate conditions (Koop 1987). *Acer argutum* often propagates asexually via suckers, while *C. japonicum* and *Acer carpinifolium* both produce suckers at the base of their main stems; these changes lead to trunk modification on long- and short-term timescales, respectively. Suckering at the base of the trunk is likely a response to the need to balance photosynthesis and respiration under a dense canopy. Furthermore,

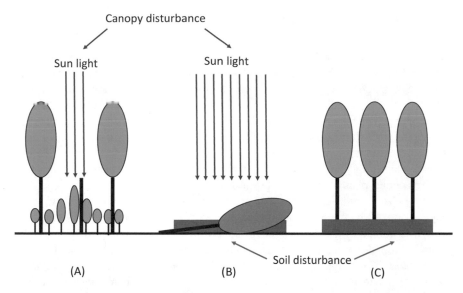

Fig. 12.1 Disturbance processes in riparian areas. Type A reflects canopy gap formation that alters light conditions. Type B reflects large-scale debris flows and landslides that alter both light and soil conditions. Type C reflects sediment flow that removes understory vegetation and alters soil conditions

individuals of the *P. rhoifolia* species are maintained via suckering in areas with heavy snowfall (Nakano and Sakio 2018).

The studies conducted at Ooyamazawa revealed that the regeneration of these species is strongly related to disturbance regimes in riparian zones. Major disturbances typical of mountain basins include predictable flooding and typhoons in the rainy season, as well as sporadic debris flows caused by large typhoons. Large-scale landslides also occur at low frequency, but with a high degree of unpredictability. Three disturbance processes were identified within the riparian zone (Fig. 12.1): (1) gap formation due to withering and truck breakage in canopy trees, which increases light availability in the understory (Fig. 12.1A); (2) small-scale sediment movement associated with mountain streams, which alters the soil environment without altering light conditions (Fig. 12.1C); and (3) large-scale landslides and debris flows, which significantly alter the soil, vegetation, and light conditions (Fig. 12.1B). The seven riparian tree species studied at Ooyamazawa were closely associated with disturbance events and environmental changes typical of mountain basins.

12.3   Plant Species Coexistence

Chapter 6 discusses the life histories of herbaceous plant species with respect to topography. In 1983, 76 species were identified in the herbaceous layer (Sakio et al. 2013). Six landform types were identified along the Ooyamazawa basin valley: debris flow terrace, alluvial fan, terrace scarp, new landslide site, old landslide slope, and talus. Herbaceous plant diversity was associated with mosaic surface soil conditions, which are a product of disturbances such as debris flows, landslides, and soil erosion. A strong association was found between micro-scale soil hetero-geneity (caused by disturbance) and the growth of herbaceous plants.

Herbaceous plants may adapt to environmental conditions through vegetative growth, as well as through sexual and asexual reproduction. At Ooyamazawa, three major groups of herbaceous plants were defined. The first were perennial herbs and ferns with rhizomes growing on stable debris flow terraces and alluvial fans. This group included spring ephemerals, such as *Corydalis lineariloba* and *Allium monanthum*. The remaining two groups included annual species (e.g., *Impatiens noli-tangere* and *Persicaria debilis*) and two subgroups of perennials: those with bulbils (e.g., *Elatostema umbellatum* var. *majus* and *Laportea bulbifera*) and those with replacement rhizomes (e.g., *Chrysosplenium macrostemon*, *Cacalia delphiniifolia*, and *Cacalia farfaraefolia*). The annual species and perennials groups tended to dominate in areas with frequent annual disturbance, such as sandbars and new landslide sites with unstable soils.

Coexistence mechanisms were investigated among the dominant canopy species, *F. platypoda*, *P. rhoifolia*, and *C. japonicum*, based on life-history characteristics and disturbance regimes (Chap. 7). *F. platypoda* was the dominant species at Ooyamazawa, comprising >60% of all canopy trees. Seedlings and saplings were common in the understory; peaks were observed for small seedlings and those with a diameter at breast height between 40 and 50 cm. *F. platypoda* has high shade tolerance and many saplings of this species were found in the understory. Further-more, mature individuals were found to have regenerated following a major distur-bance event at Ooyamazawa, approximately 200 years ago. Seedlings and saplings typically have greater resistance to flooding; therefore, they can survive at stream edges and grow rapidly to reach canopy height when gaps are formed.

Canopy-layer, mature individuals of the *P. rhoifolia* species formed a large patch approximately 50 m in diameter. Presumably, these individuals were approximately the same age and had regenerated following a past disturbance event. *P. rhoifolia* exhibited a faster growth rate under direct sunlight relative to other species, a trait that was also observed in a nursery experiment. However, its lifespan is relatively short, with a maximum of 150 years; thus, mature individuals are replaced by other species after death.

C. japonicum displayed a relatively uniform age and diameter at breast height size class distribution at Ooyamazawa. Larger individuals were clustered near the moun-tain stream, while seedlings and saplings were uncommon in this area. This indicated limited regeneration opportunity for *C. japonicum* at Ooyamazawa. Sub-canopy layer individuals of *C. japonicum* were located within the regenerated stand of *P. rhoifolia*, which suggested that the regeneration potential of *C. japonicum* also relies on large-scale disturbance. *C. japonicum* typically has high annual seed production and its seeds are dispersed over long distances (Sato et al. 2006), which allows for the colonization of newly disturbed areas. Once established, individuals are maintained by suckering around the main trunk. Coexistence mechanisms in the riparian forest at Ooyamazawa reflected niche partitioning at early stages; however, long-term coexistence relied on unpredictable large-scale disturbance events. These results reflect a similar trend observed in the understory herbaceous species at the site.

The rough topography at Ooyamazawa is the result of an earthquake that occurred 200 years ago (Sakio 1997) and a massive landslide that occurred 100 years ago (Sakio et al. 2002). More recently, climate change has caused changes in the disturbance regime, especially in the riparian area. The frequency and magnitude of large-scale disturbance events has increased, exemplified by typhoon Hagibis (Typhoon No. 19) in 2019. These changes may affect long-term patterns in plant regeneration and life-history strategies at the site.

12.4 Effects of Global Warming on Tree Reproduction

Global warming has progressed over the past century and has been related to various ecosystem-level changes. The effects of global warming will first become evident in polar and high-altitude ecosystems (Grabherr et al. 1994; Kullman 2001; Sanz-Elorza et al. 2003; Sturm et al. 2001; Wardle and Coleman 1992). Grabherr et al. (1994) demonstrated significant ecological impacts of global warming in terms of the upwards advance of alpine-nival flora. The timberline of Mt. Fuji advanced rapidly upwards from 1978 to 1999 (Sakio and Masuzawa 2011). Furthermore, Kudo and Suzuki (2003) showed accelerated growth in a few restricted species using artificial warming over a 5-year period; they also showed a change in vegetation structure in a mid-latitude alpine ecosystem. The maximum, average, and minimum temperatures in Chichibu increased by 1.09 °C, 1.15 °C, and 1.69 °C, respectively, from 1926 to 2018 (Fig. 12.2).

Phenological change was documented in *F. platypoda* at Ooyamazawa over a 28-year survey period (Chap. 2). The flowering interval of *F. platypoda* was 2–3 years for both female and male trees, and flowering of the two sexes showed clear synchronization, until 2002. After 2002, an increasing number of males flowered annually, while females retained a flowering interval, and synchronization was lost. The reason for this change is unclear, but is presumably related to an increase in the photosynthetic rate due to rising temperature and an increase in net production due to a prolonged photosynthetic period.

In the past, significant changes in the distribution of forest vegetation have been recorded in the Japanese archipelago due to climate change, such as during the glacial and interglacial periods. In many instances, species' ranges may shift due to

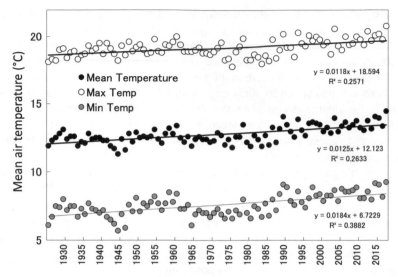

Fig. 12.2 Maximum, average, and minimum temperatures in Chichibu city from 1926 to 2018 (Japan Meteorological Agency 2013)

the direct physiological effect of temperature; however, there may also be instances in which local populations are extirpated due to unsuccessful reproduction.

12.5 Sika Deer Browsing and Riparian Forest Ecosystems

12.5.1 Understory Vegetation and Tree Species

LTER documented an unexpected phenomenon at Ooyamazawa (Chap. 8). The sika deer population in Japan has nearly doubled over the past few decades (Miura and Tokida 2008). After 2000, herbaceous vegetation cover sharply declined at the site (Sakio et al. 2013); in particular, vegetation cover quickly decreased to only a few percent and species richness was reduced by half. The majority of herbaceous species were affected by deer browsing, excluding the toxic *Aconitum sanyoense*, *Scopolia japonica*, and *Veratrum album* ssp. *oxysepalum*. Small spring ephemeral species were also less affected due to their short growing season.

Sika deer browsing further influenced tree species populations at Ooyamazawa. Many trees, including canopy-layer individuals, died due to girdling by deer. This damage was species-specific, wherein *Ulmus laciniata* exhibited the greatest damage. *Acer* spp., especially *Acer carpinifolium*, were also extensively damaged. Browsing on suckers of *Acer carpinifolium* was sufficiently extensive to cause the death of some individuals. To restore the vegetation by exclusion of deer predation, fences were installed in the Ooyamazawa basin, excluding the riparian area, in 2016 (Fig. 12.3). Long-term monitoring will be critical to determine the effectiveness of the fencing.

Fig. 12.3 Exclusion
fencing for sika deer at
Ooyamazawa. Fencing was
established in 2016

Fig. 12.4 Decline in a Sasa
borealis community on a
hillside slope. This decline
was a result of sika deer
browsing and simultaneous
flowering

12.5.2   Avifauna

Within the Ooyamazawa riparian forest research site, 44 species were recorded
during the breeding season and 23 were recorded in the wintering season between
2010 and 2017. An 8-year survey that began in 2010 suggested that the avifauna
community dramatically changed at this site over time.

Sika deer browsing (Chap. 8) led to the decline of the *Sasa borealis* community
on hillside slopes (Fig. 12.4), which has not yet recovered since blooming in 2013.

The abundance of bird species that feed in shrubs, namely *Cettia diphone*, *Luscinia cyane*, *Luscinia akahige*, and *Leiothrix lutea*, significantly decreased over the 8-year period; therefore, these species were nearly absent by 2017. Furthermore, the abundance of bird species that nest in shrubs, such as *Phylloscopus borealoides* and *Phylloscopus coronatus*, also decreased.

12.5.3 Ground Beetle Community

Ground beetle monitoring at Ooyamazawa, conducted from 2008 to 2017, captured 2381 individuals from 19 families (including 1969 individuals of 36 carabid species). The carabid community was characterized by high species richness and a high proportion of Japanese endemic species, relative to other forest monitoring sites across Japan. Similar to the avifauna, the ground beetle community showed major changes over the course of the survey period. Most carabid species exhibited dramatic declines and the annual catch of carabids decreased by 80% over the sampling period. While some of this decline was related to the decline in understory vegetation due to deer browsing, it may also have been related to climate change and warming.

12.5.4 Cascade Effects of Sika Deer Browsing

Sika deer alter the structure and composition of forests through herbivory. Consequently, regeneration is prevented (Takatsuki 2009). The sika deer population increase in Japan has resulted from deer protection policies, increased food resources related to establishment of plantations after World War II, reduced hunting pressure, and reduced snow cover in winter. Cascade effects that result from deer browsing are described in Fig. 12.5. At Ooyamazawa, sika deer browsing led to a reduction in forest floor vegetation (Chap. 8), alterations in beetle and bird communities (Chaps. 10 and 11), and a decline in tree populations due to girdling (Chap. 9).

Although not addressed in this study, the reduction in understory vegetation would be expected to increase surface soil erosion, leading to deterioration in stream water quality and detrimental effects on aquatic life, particularly breeding fish (Sakai 2013). Previous research suggested that the increase in deer population affected fish populations in the upper reach of the Yura River, located in the Ashiu Forest Research Station of the Kyoto University Field Science Education and Research Center, Kyoto Prefecture, Japan (Nakagawa 2019).

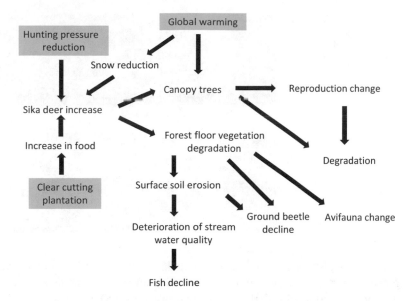

Fig. 12.5 Cascade effects on riparian ecosystems caused by human impacts and global change

12.6   Conservation of Riparian Forest Ecosystems

Following World War II, dams were constructed in mountain basins in Japan (Fig. 12.6). Additionally, plantations of *Cryptomeria japonica* and *Chamaecyparis obtusa* were established, often directly adjacent to riparian zones (Fig. 12.7). As a result, many riparian forests were lost, and mountain stream ecosystems were degraded (Sakio and Suzuki 1997). Given this legacy, conservation of the remaining areas of intact riparian forest is critical. These forests can become models for conservation   and   rehabilitation,   while   serving   as   important   genetic

Fig. 12.6 Erosion control dams in mountain streams

Fig. 12.7 *Cryptomeria japonica* plantation in a riparian zone following clear cutting of native forest

resources. Although sika deer pose an ongoing threat, deer management and exclusion fencing efforts are beginning to show positive effects in some areas in Japan. Deer fencing was installed in 2016 at Ooyamazawa (Chap. 9), but recovery has not yet been observed. Future monitoring efforts will determine the feasibility of understory vegetation restoration. Existing long-term survey data will allow for comparison between recovering exclusion areas and the vegetation community composition prior to intensive deer browsing.

12.7 Future Directions

The research conducted at Ooyamazawa highlights the importance and value of LTER. The initial research objectives included understanding life-history characteristics in riparian forests and determining long-term forest dynamics. However, deer browsing intensified after the establishment of the site and provided an excellent opportunity to study the influence of browsing on forest dynamics. Existing vegetation survey data allowed us to directly assess changes in vegetation caused by sika deer (Chaps. 8 and 9). Effects on avifauna and ground beetle communities were also assessed. Our 28-year analysis of *F. platypoda* clearly demonstrated changes in flowering and seed production resulting from climate change and rising temperatures.

LTER requires consistent research effort and continuous funding. Thus far, LTER has been conducted at survey sites throughout Japan and produced important findings. We note that researchers who are invested in LTER from its conception often remain highly motivated; however, wardens may see routine monitoring as burdensome. The continuation of manipulative experiments and long-term monitoring is an important concern for the future. Fortunately, research at Ooyamazawa has benefited from the addition of young researchers, which is reflected in the efforts of

JaLTER and the Monitoring Sites 1000 Project. We look forward to continued research efforts.

References

Grabherr G, Gottfried M, Pauli H (1994) Climate effects on mountain plants. Nature 369:448

Japan Meteorological Agency (2013). http://www.data.jma.go.jp/obd/stats/etrn/index.php. Accessed 5 Nov 2019

Koop H (1987) Vegetative reproduction of trees in some European natural forests. Vegetatio 72:103–110

Kudo G, Suzuki S (2003) Warming effects on growth, production, and vegetation structure of alpine shrubs: a five-year experiment in northern Japan. Oecologia 135:280–287

Kullman L (2001) 20th century climate warming and tree-limit rise in the southern Scandes of Sweden. Ambio 30:72–80

Miura S, Tokida K (2008) Management strategy of sika deer based on sensitivity analysis. In: McCullough DR, Kaji K, Takatsuki S (eds) Sika deer: biology and management of native and introduced populations. Springer, Tokyo, pp 453–472

Nakagawa H (2019) Habitat changes and population dynamics of fishes in a stream with forest floor degradation due to deer overconsumption in its catchment area. Conserv Sci Pract 1:e71

Nakano Y, Sakio H (2018) The regeneration mechanisms of a *Pterocarya rhoifolia* population in a heavy snowfall region of Japan. Plant Ecol 219(12):1387–1398

Nakashizuka T, Matsumoto Y (eds) (2002) Diversity and interaction in a temperate forest community: Ogawa Forest Reserve of Japan. Springer, Tokyo

Sakai M (2013) Changes in interaction between forest-stream ecosystems due to deer overbrowsing. J Jpn Soc Reveget Tech 39:248–255

Sakio H (1997) Effects of natural disturbance on the regeneration of riparian forests in a Chichibu Mountains, central Japan. Plant Ecol 132:181–195

Sakio H, Masuzawa T (2011) The advancing timberline on Mt Fuji: natural recovery or climate change? J Plant Res 125:539–546

Sakio H, Suzuki W (1997) Overview of riparian vegetation: structure, ecological function and effect of erosion control works. J Jpn Soc Erosion Control Eng 49:40–48

Sakio H, Kubo M, Shimano K, Ohno K (2002) Coexistence of three canopy tree species in a riparian forest in the Chichibu Mountains, central Japan. Folia Geobot 37:45–61

Sakio H, Kubo M, Kawanishi M, Higa M (2013) Effects of deer feeding on forest floor vegetation in the Chichibu Mountains, Japan. J Jpn Soc Reveget Tech 39:226–231

Sanz-Elorza M, Dana ED, González A, Sobrino E (2003) Changes in the high-mountain vegetation of the Central Iberian Peninsula as a probable sign of global warming. Ann Bot 92:273–280

Sato T, Isagi Y, Sakio H, Osumi K, Goto S (2006) Effect of gene flow on spatial genetic structure in the riparian canopy tree *Cercidiphyllum japonicum* revealed by microsatellite analysis. Heredity 96:79–84

Sturm M, Racine C, Tape K (2001) Increasing shrub abundance in the Arctic. Nature 411:546–547

Takatsuki S (2009) Effects of sika deer on vegetation in Japan: a review. Biol Conserv 142:1922–1929

Wardle P, Coleman MC (1992) Evidence for rising upper limits of four native New Zealand forest trees. N Z J Bot 30:303–314

Index

A

Abandoned channel, 12, 31
Abies, 140
Abies homolepis, 16, 144, 145, 167
Abortion, 28
Accipiter nisus, 203
Acer amoenum, 146
Acer amoenum Carrière var. *amoenum*, 84, 167
Acer amoenum var. *amoenum*, 89
Acer amoenum var. *moenum*, 88
Acer argutum, 13, 84–91, 93–95, 146, 165, 167, 169–171, 174–176, 216
Acer carpinifolium, 13, 84–91, 93–95, 145, 147, 170, 171, 174, 175, 216, 220
Acer cissifolium, 84, 88, 89, 168
Acer distylum, 84, 88, 89, 168
Acer japonicum, 167
Acer maximowiczianum, 84, 88, 89, 167
Acer micranthum, 84, 88, 89, 146
Acer mono, 52, 145
Acer nipponicum, 84, 88, 89, 145, 167
Acer palmatum, 84, 88, 89, 167
Acer pictim, 84–86, 88–91
Acer pictum, 13, 84, 87, 93–95, 165, 167, 170, 171, 174
Acer rufinerve, 84, 88, 89, 145, 167
Acer shirasawanum, 13, 59, 84–91, 93–95, 146, 165, 167, 169–171, 174
Acer spp., 16, 59, 83–95, 216, 220
Acer tenuifolium, 84, 88, 89, 167
Acer tschonoskii, 168
Aconitum loczyanum, 146
Aconitum sanyoense, 105, 144, 145, 153, 220
Aconitum tonense, 169, 174
Actinidia arguta, 70, 146, 167

Active channel, 12, 31
Adoxa moschatellina, 105, 144, 145, 148, 153, 154
Advanced sapling, 34, 35
Advanced seedling, 131, 133
Adventitious root, 25
Aegithalos caudatus, 203
Aesculus turbinata, 40, 59
Age of reproductive maturity, 64
Air humidity, 9
Air temperature, 7
Allium monanthum, 104, 105, 108, 154, 156, 218
Alluvial fan, 218
Alnus hirsute, 52
Alpine zone, 187
Androdioecy, 26
Anemonopsis macrophylla, 146
Anemophilous species, 41
Angelica polymorpha, 146, 157
Annual seed production fluctuates, 42
Annual species, 218
Anther, 26
Antler fraying, 164
Apus pacificus, 203
Aria alnifolia., 167
Arisaema ovale var. *sadoense*, 146
Arisaema tosaense, 146
Asarum caulescens, 13, 105, 144, 145, 150
Ash, 24
Asian continent, 188
Astilbe thunbergii, 105
Astilbe thunbergii var. *thunbergii*, 146, 148
Athyrium wardii., 145
Avifauna, 216, 221, 222, 224

Printed in the United States
By Bookmasters